SYSTEM MATERIALS
BLDG. 250

The sulphate-reducing bacteria

The sulphate-reducing bacteria

Second edition

J. R. POSTGATE, FRS

PROFESSOR OF MICROBIOLOGY
UNIVERSITY OF SUSSEX

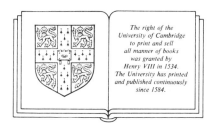

CAMBRIDGE UNIVERSITY PRESS
CAMBRIDGE
LONDON   NEW YORK   NEW ROCHELLE
MELBOURNE   SYDNEY

Published by the Press Syndicate of the University of Cambridge
The Pitt Building, Trumpington Street, Cambridge CB2 1RP
32 East 57th Street, New York, NY 10022, USA
296 Beaconsfield Parade, Middle Park, Melbourne 3206, Australia

© Cambridge University Press 1979, 1984

First published 1979
Reprinted 1981
Second edition 1984

Printed in Great Britain by the University Press, Cambridge

Library of Congress catalogue card number: 83-15307

British Library cataloguing in publication data
Postgate, J. R.
The sulphate-reducing bacteria — 2nd ed.
1. Sulphur bacteria
I. Title
589.9   QR92.S8

ISBN 0 521 25791 3
(ISBN 0 521 22188 9 First edition)

UP

Contents

		Page
Preface to the Second Edition		vii
Preface to the First Edition		ix
1	Introduction	1
2	Classification	9
3	Cultivation and growth	30
4	Structure and chemical composition	51
5	Metabolism	56
	Some theoretical data	56
	Broad metabolic patterns	59
	CO_2 assimilation	61
	Carbon dissimilation	62
	Hydrogen metabolism	70
	Electron transport	73
	Dissimilatory sulphur metabolism	82
	Energy metabolism	91
	Metabolism of nitrogen	96
	Metabolism of phosphorus	98
	Metabolism of iron	98
	Metabolic regulation	99
	Metabolic inhibitors	99
6	Evolution	101
7	Ecology and Distribution	107
	Environmental effects of sulphate reduction	110
	Populations in natural environments	112
	Consortia and syntrophs	114
	Role in the sulfuretum	116

8	Economic activities	124
	Pollution of water	124
	Release of phosphate in water	129
	Pollution of soil and sand	129
	Purification of wastes	130
	Corrosion of metals and stonework	132
	Formation of mineral deposits	138
	Food spoilage	144
	Oil technology	145
	Contamination of gas	150
	Paper technology	150
	Animal nutrition	150
	Pathogenicity	151
	Miscellanous economic activities	151
9	Epilogue	153
	Appendix 1: Characters of certain strains of sulphate-reducing bacteria held in National collections of bacteria	155
	Appendix 2: Selected list of inhibitors of sulphate-reducing bacteria	163
	References	169
	Index	200

Preface to the Second Edition

There is an element of irony in the fact that, even as I was writing the first edition of this book, knowledge of the sulphate-reducing bacteria was undergoing a revolution. A hint appeared on p. 23 of that edition, where I alluded to the announcement by Professor Pfennig at an International Congress in 1978, of the discovery of some new species of sulphate-reducing bacteria. But I, and others, little realized the new vistas which Professor Pfennig and his colleague Dr Widdel were opening with their beautiful microbiology: a positive menagerie of filamentous, lemon-shaped, coccoid, egg-shaped, curved and rod-shaped sulphate-reducing bacteria emerged, many able to utilize an unsuspected variety of substrates. The utilization of acetate, hitherto doubted, became commonplace. Having been one of the major doubters myself, I fully appreciated the courteous irony which prompted Norbert Pfennig and Fritz Widdel to name for me the sulphate reducer which uses acetate but almost nothing else ...[1] I am honoured, gentlemen, and only slightly repentant. The existence and properties of these new organisms had, and are still having, profound effects on our ideas about the classification, ecology, evolution and physiology of this group. It became obvious that a radical revision of my monograph was becoming obligatory.

So here it is. The biochemistry of the older organisms, and of some of the new ones, had meanwhile been progressing and it had also become clear to me that the section on economic activities was too short: I had left too much out because it was available in previous reviews and this policy (which arose not from laziness but from a reluctance to repeat myself) was a nuisance to readers. So I have taken the opportunity to revise and extend every chapter in this book, even to the epilogue and appendices. The bibliography is expanded, though still far from exhaustive, for this is a monograph, not a treatise, and I have had to omit reference to many

---

[1] *Desulfobacter postgatei* (Widdel & Pfennig, 1981a).

important primary publications, particularly when they were discussed in a reasonably accessible review. However, those who rightly complained may notice that titles of papers are now included in the list.

Professor Pfennig and Dr Widdel have been most helpful in providing me with photographs and information for this revision and I thank them very much. I have also corrected some errors – happily rather few – which blemished the first edition and sought criticisms from various authorities; I am grateful to Bo Jørgensen, David Nedwell, Rivers Singleton, Graham Skyring, Robert Starkey and Claude ZoBell in particular for their helpful comments, but omissions and errors of fact and/or interpretation are still my fault.

Acknowledgements. I thank Mrs Brenda Hall for typing the manuscript and my wife for helping with the text and references. Others who kindly provided figures or information are acknowledged in the text.

January 1983   John Postgate

Preface to the First Edition

The sulphate-reducing bacteria are, as I hope this monograph will demonstrate, a bizarre group of microbes of which most people, including many microbiologists, know nothing. Yet these organisms impinge on our lives in a variety of subtle, and occasionally blatant, ways. Despite their fascinating qualities they have been a somewhat neglected backwater of microbiological research: smelly, awkward to grow, intractable to isolate and count, but revealing intriguing novelties of biochemistry and physiology to those persistent enough to stick with them. The late K. R. Butlin pointed out that the Dutch, whose canals often provide so generously foetid a habitat for these bacteria, had a vested interest in knowing about them, and it is no coincidence that they were discovered by the great Dutch microbiologist M. W. Beijerinck, nor that Dutchmen such as Elion, van Delden, Baars and Kluyver laid the groundwork of today's knowledge. Butlin numbered determination among his many qualities and I owe him a lifelong debt for introducing me to these bacteria in the late 1940s, at an early stage in my scientific career, pointing my nose in the right direction and leaving me to get on with it. In due course I came to know personally nearly everyone in the world working on these bacteria: Bill Bunker, Claude ZoBell, Robert Starkey, Syd Rittenberg, Jacques Senez, Leon Campbell, Harry Peck, Jean Le Gall and a small host of others – a few I knew only by correspondence. It was a small, friendly scientific community within which rivalries and antagonisms, while not completely absent, played no important part in the accumulation and distribution of scientific information. Today, when the struggle for priority in publication has made much of scientific research a disagreeable rat-race, I recall our earlier academic calm with perhaps rose-tinted nostalgia. For there were jealousies and unseemly rushes into print, but when we met we still discussed our work freely and often exchanged manuscripts before submitting them for publication (a practice which is certainly rare among today's scientists, at least in 'trendy' areas of research). The sulphate-reducing bacteria never really became trendy, though occasional catastrophes (a corrosion disaster, a world sulphur shortage) or a spectacular discovery (a cytochrome structure, their mixotrophy) might lead to a brief display of their talents in the popular or serious scientific press.

Inconspicuousness has its advantages, and I hope this monograph will not bring sulphate reduction too far into the forefront of competitive research. On the other hand, these are very important microbes, not only from an academic point of view but also in numerous practical ways. And it is regrettable that the student, teacher or technologist, if he seeks to find out about them, will find them dismissed in a paragraph or two of most microbiological textbooks and will have to burrow into quite obscure and ancient reviews of topics ranging from metallurgy to straight microbiology. In this monograph I have tried to compensate for that situation: to pull together in compact form the state of our knowledge of these bacteria as we approach the 1980s. I have also tried to write down some of the microbiological 'lore' necessary for handling these bacteria, such as that which gave my colleague, the late Miss M. E. Adams, green (black?) fingers, enabling her to isolate and maintain the first collection of reliably pure cultures in the world.

Microbiology is a science, but a touch of art and craft is always desirable, even essential, for progress to be made. I offer this monograph, then, as a largely but not entirely scientific handbook for those whose academic or practical compulsions have brought them face to face with these exotic forms of life.

October 1978   John Postgate

1

Introduction

Sulphur compounds are widespread on this planet and, like most of the commoner elements, the sulphur atom has become an essential component of the biosphere: part of the chemical structure of living things. Protoplasm contains between 0.4% and 1% of sulphur (as organic sulphur compounds), depending on the type of cell and the environment from which the cells came. For many millions of years – at least 5×10^8 years – the sulphur at the surface of the planet (available to the biosphere) has existed predominantly in an oxidized form: as sulphates in soils, rocks, rivers and seas; as sulphur oxides, a minor component of the atmosphere. For such sulphur to be mobilized for biological use, it must be reduced, so the biological reduction of sulphate, like biological nitrogen fixation and biological oxygen production, has become recognized as one of the critically important processes on which life on the planet depends. As far as we know, animals do not conduct this reaction: from protozoa to man, they appear to depend on plants and/or microbes for their supplies of reduced sulphur. Green plants, fungi such as yeast and many species of bacteria, unlike animals, can use sulphate as their sole source of the biological element sulphur. In so doing, they reduce the sulphate ion, bringing the sulphur atom from its fully oxidized state to its fully reduced state. This process is *assimilatory sulphate reduction*, so called because the sulphur is assimilated: it is incorporated into the organisms' protein as sulphur-containing amino acids or built into co-factors such as biotin and pantothenic acid. The biochemical pathways and means of regulation of assimilatory sulphate reduction are fairly well understood, but this subject will not form part of this monograph. Instead, I shall be concerned with a different process, of perhaps equal biological importance, conducted by a group of bacteria which, as far as we know at present, is unique in its peculiar physiology.

The name 'sulphate-reducing bacteria' is conventionally reserved for a class of microbes which conducts *dissimilatory sulphate reduction*. In this process the sulphate ion acts as an oxidizing agent for the dissimilation of organic matter, as does oxygen in conventional respiration. A small amount of reduced sulphur is assimilated by the organism, but virtually all is released into the external environment as the sulphide ion, usually substantially hydrolysed to free H_2S. The process has also been called 'sulphate respiration', analogous to 'nitrate respiration' found among nitrate-reducing and denitrifying bacteria. In addition, it has superficial analogies to carbonate reduction, a process conducted by certain methanogenic bacteria. Sulphate respiration is not encountered on this planet outside certain specialized bacteria. To provide a quantitative guide to the difference in scale between assimilatory and dissimilatory sulphate reduction, one can make the following comparison. In conditions of sulphate limitation *Klebsiella aerogenes* yields about 200 mg dry wt organisms/mg sulphur (Postgate & Hunter, 1962); the yield with *Desulfovibrio* (a group of sulphate-reducing bacteria) depends on the carbon source but is in the region of 0.5 to 1 mg dry wt organisms/mg sulphur.

Though the sulphate-reducing bacteria have been known for over seven decades, little information about them has penetrated to conventional microbiology textbooks. Therefore this monograph, though primarily concerned with information gained during the last three decades, will necessarily allude to earlier work. These bacteria have been the subjects of periodic reviews in the specialized literature; those by Bunker (1936), Starkey & Wight (1945), Starkey (1960/61), Postgate (1959*a*, 1960*a*, 1965*a*) and Le Gall & Postgate (1973) form a series which may be consulted for amplification of various aspects mentioned in this monograph.

The sulphate-reducing bacteria were discovered by Beijerinck (1895); van Delden (1903) reported marine, salt-tolerant varieties and Elion (1925) described thermophilic types. Baars (1930), in a thesis which was published but which is not widely available, provided a most extensive study of these bacteria, one which is still an absorbing document though much of its information has been superseded. Such early work was reviewed by Bunker (1936) and briefly by Butlin, Adams & Thomas (1949). The sulphate-reducing bacteria are all very strict anaerobes. Some are now known to be capable of fermentative growth in the absence of sulphate, analogous to the fermentative growth of a yeast without oxygen, but none can grow with oxygen as electron acceptor, and oxygen always inhibits their growth. They grow relatively slowly compared with a common soil or water organism such as *Pseudomonas* (partly because growth of cultures is often non-exponential, see Chapter 3) but they have a remarkable capacity for survival in terrestrial and aquatic environments (see Chapter 7). They are widely distributed, ready to become active whenever local conditions become anaerobic.

They play an important part in the biological sulphur cycle and this must be discussed briefly before proceeding further.

The sulphur cycle

Sulphur, like all the biological elements, is transformed and translocated in the biosphere by a combination of biological and chemical agencies. Burning of fuels, particularly fossil fuels, liberates sulphur oxides into the atmosphere (causing a pollution problem familiar to environmentalists); volcanic and thermal emissions release both sulphur oxides and hydrogen sulphide (with some sulphur vapour) into the atmosphere. The sulphur compounds are deposited on trees, buildings and stonework as sulphurous and sulphuric acid, or are washed as sulphate by rain into soil. From such sites they leach into rivers and, ultimately, the sea. Sea water and many types of land waters contain sulphates, and spray from such waters leads to formation of aerosols which translocate sulphur about the biosphere. These processes are, broadly speaking, physico-chemical ways in which the S atom is turned over in the biosphere. They are supplemented by the biological processes illustrated in Fig. 1, which is a formalized scheme of the chemical transformations undergone by the sulphur atom in nature through biological agencies, analogous to the rather better known biological nitrogen cycle. Chemical and biological processes can be combined in geochemical sulphur cycles (e.g. Kellog et al., 1972; Granat, Hallberg & Rodhe, 1976), which describe the translocation and the transformations of sulphur in the biosphere by both chemical and biological agents. Some 10^8 tonnes S pass through the atmosphere in a year, and about 50% of this may be biogenic: released by sulphate reduction and decomposition of organic sulphur. The microbiology of the sulphur cycle has recently undergone some radical changes, notably the discovery of the sulphur-reducing bacteria; some of these (e.g. *Desulfurococcus*; Zillig et al., 1982) are remarkably acid-tolerant, extremely thermophilic archebacteria. The microbiology of the conventional sulphur-reducing bacteria was discussed by Pfennig & Biebl (1981), to which article the reader is referred. The sulphate-reducing bacteria contribute substantially to the sulphur cycle; as is illustrated in the biological version presented in Fig. 1, their main role is to bypass assimilatory sulphate reduction and generate H_2S. They do this in sufficient amounts to support growth of the sulphide- and sulphur-oxidizing bacteria and thus they can generate a microbial ecosystem consisting of interdependent sulphur-oxidizing and sulphate-reducing bacteria which is called a 'sulfuretum'. Their activities, either direct or through the sulphur cycle, have a variety of ecological and economic consequences; their role in the sulfuretum is discussed in more detail in Chapter 7 (p. 116).

Some historical errors

Because of the strict anaerobic habit and slow growth of sulphate-reducing bacteria, many of the earlier workers used impure cultures, albeit unintentionally; even in 1949 it seemed likely that only a few pure cultures were available in the world (Butlin *et al.*, 1949) and not all those were as

Fig. 1. The biological sulphur cycle. Sulphur (SO_4^{2-}) is reduced to sulphide (S^{2-}) by dissimilatory sulphate-reducing bacteria and provides substrates for sulphide-oxidizing bacteria which convert it, by way of elemental sulphur (S^0) back to sulphate. In assimilatory sulphate reduction, the sulphur of sulphate passes through the sulphide level of oxidation and becomes incorporated into an amino acid (RSH) before being built into plant as microbial protein. This is eaten by animals and the sulphur is eventually returned to the cycle as sulphide formed during the breakdown and putrefaction (by bacteria) of dead organisms.

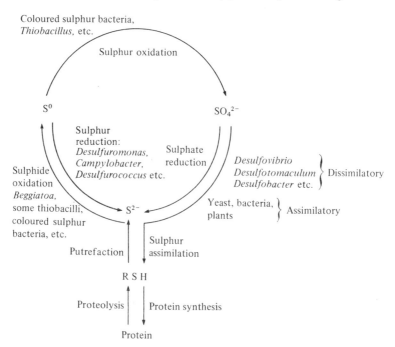

Introduction

pure as their proprietors believed (Postgate, 1953a). New methods for obtaining pure cultures are now available which, together with more explicit criteria of purity, have somewhat eased the problem of contamination (see Chapter 3), but a few instances in which misinformation arose, attributable at least in part to the use of impure cultures, should be mentioned.

Desulfovibrio rubentschickii. Pure cultures of most sulphate-reducing bacteria, when utilizing ordinary carbon substrates with more than three carbon atoms per molecule (e.g. lactate or malate), dissimilate the carbon source to only the acetate level of oxidation; acetate accumulates and is not a growth substrate. Yet in crude enrichment cultures acetate slowly disappears and in nature acetate does not accumulate even where sulphate-reducing bacteria are very active. An acetate-utilizing species, *Desulfovibrio rubentschickii*, was described by Baars (1930). Exhaustive attempts to re-isolate this species failed in several laboratories, though ordinary acetate-forming sulphate-reducing bacteria were easily detected, which suggested that *D. rubentschickii* must have been some kind of mixed commensal population (see Selwyn & Postgate, 1959). There the position rested for several years, with no further clue as to where the acetate went, until the mid-1970s, when Pfennig & Biebl (1976) discovered a 'sulphur-reducing bacterium', *Desulfuromonas acetoxidans*, which reduced the element sulphur (but not sulphate or sulphite) to sulphide at the expense of acetate.

$$4S + NaOOC.CH_3 + 3H_2O \rightarrow 4H_2S + NaHCO_3 + CO_2$$
$$(\Delta G = -23.8 \text{ kJ})^1$$

This organism provided a partial explanation of the disappearance of acetate: its oxidation with partly reduced sulphur could occur. However, a year later Widdel & Pfennig (1977) solved the problem in principle by isolating a true acetate-oxidizing sulphate-reducing bacterium of the spore-forming group (*Desulfotomaculum*). They named it *Desulfotomaculum acetoxidans*; contrary to Baars's report, it does not utilize the common substrates of *Desulfovibrio* (such as lactate and pyruvate) and it is also distinctive in other ways. Subsequently they isolated new genera of

---

[1] These values have been converted from calories to Joules because the latter are now the International System (SI) units of energy. For comparison with data published earlier, 1 cal = 4.18 J.

acetate-oxidizing, sulphate-reducing bacteria (Widdel, 1980), mostly different in several respects, including morphology, from *Desulfovibrio*. These organisms are discussed in Chapter 3. Whether Baars and later workers actually had such species in their cultures will probably never be known, but the work of Pfennig and Widdel showed satisfactorily that acetate-supported sulphate reduction can and does occur and is the property of distinct species and genera.

Interconversion of mesophilic and thermophilic types. Sulphate-reducing bacteria comprise both mesophilic strains which grow best around 30 °C but tolerate up to about 42 ° and thermophilic strains able to grow at temperatures between 50 and 70 °C. Kluyver & Baars (1932) believed that these were adaptive variants of the same organism, a view supported by Starkey (1938) and other workers including H. J. Bunker (see Postgate, 1953*b*). However, Campbell, Frank & Hall (1956) showed conclusively that the thermophilic types as commonly isolated, and available from collections, were a completely different species, hitherto known as *Clostridium nigrificans* and not recognized as sulphate-reducing bacteria. This finding ultimately led to their reclassification in the genus *Desulfotomaculum* (see Chapter 3). My unpublished experiments in 1954–56 also tended to support the views of Campbell *et al.* (1956) and I was later fortunate in being able to inspect the laboratory notes from 1936–38 of the late H. J. Bunker, when he believed he had substantiated Baars's and Starkey's findings. On close examination, it was clear that the viability of the mesophilic (30 °C) strains cultured at 55 °C had not been checked: Bunker had only given them one passage at the 'thermophilic' temperature. He used cultures of about 60 ml volume, and blackening, due to FeS precipitation (see Chapter 3), was his criterion of growth. It was a matter of simple experiment to show, in 1958, that a vigorous culture of mesophilic desulfovibrios could undergo one division, perhaps two, *while a 60-ml culture was warming up* in a conventional 55 °C incubator. Certainly it could produce enough H_2S to simulate growth during that period. Bunker's mesophiles which had 'adapted' to thermophily were not rigorously checked for acquirement of the thermophilic character – reasonably enough in view of the published background. His thermophilic species which had 'adapted' to being a mesophilic species still grew very slowly – and this is true of most strains of *Desulfotomaculum nigrificans*. These criticisms do not apply to earlier work and it is less easy to account for the observations of Baars and Starkey; in discussing the reasons for those earlier findings, Campbell & Postgate (1965) could only conclude that the populations which apparently showed convertibility from mesophilic to thermophilic types were initially mixtures of both types.

Today, the adaptive interconversion of mesophilic and thermophilic species of sulphate-reducing bacteria must be regarded as mistaken. This view does not preclude the existence of naturally-occurring strains of unusual temperature habit and, indeed, natural isolates of both mesophilic *Desulfotomaculum* and thermophilic *Desulfovibrio* have been reported (see Chapter 3).

Presence of syntrophic contaminants. Postgate (1953*a*) was obliged to revise a number of quantitative data covering the biochemistry of a strain because it proved to be contaminated with a non-sulphate-reducing organism. This organism was a strict anaerobe and required the presence of the sulphate reducer for growth, except in very rich media. Thus it eluded the then customary methods of checking for contaminants.

Mixotrophy. A fourth historical error did not arise from impure cultures but is nonetheless important. It concerns their status as autotrophs. Most strains of *Desulfovibrio*, and of some other genera, can use gaseous hydrogen for the reduction of sulphate:

$$4H_2 + SO_4^{2-} \rightarrow 4H_2O + S^{2-}$$

If the energy yield of this reaction could be coupled to the assimilation of CO_2, the organisms would be capable of growing in a purely mineral environment: they would be true autotrophs. Starkey & Wight (1945) believed that this was so, although they were not certain that the cultures they used were pure. Butlin & Adams (1947) thought they had evidence for weak but real autotrophic growth of pure cultures of *Desulfovibrio* because more bacteria grew in a mineral medium under H_2 than under N_2. Adams, Butlin, Hollands & Postgate (1951) isolated a hydrogenase-deficient variant strain which did not show improved growth under H_2. However, tests with labelled CO_2 a decade later failed to confirm autotrophy (Mechalas & Rittenberg, 1960; Postgate, 1960*b*) because the proportion of cell carbon derived from CO_2 was only a few per cent more than was observed with a typical heterotroph but quite insufficient for an autotroph. Mechalas & Rittenberg (1960) concluded that the apparent autotrophy was not real. It transpired that assimilation of organic impurities in the putatively mineral media was being stimulated by H_2. This process could be duplicated by substrates such as *iso*-butanol, which acted as a source of H_2 but was not itself assimilated. Today desulfovibrios are recognized not to be true autotrophs, though they are capable of a coupled assimilaton of acetate and CO_2 together, a reaction – termed mixotrophy – which verges upon autotrophy (see Chapter 3).

Widdel (1980) reported that *Desulfonema* and *Desulfosarcina* strains could grow as true autotrophs, a claim which, at the time of writing, requires confirmation by measurements of incorporation of labelled CO_2.

Guanine : cytosine ratios. An error in some early calculations of the base composition of DNA of sulphate-reducing bacteria, a parameter of taxonomic importance (p. 22), was adumbrated by Skyring & Jones (1972).

Enumeration. Finally, while discussing historical errors, the fact should be mentioned that all viable counts of sulphate-reducing bacteria reported before 1955 were probably incorrect, as well as many reported after that date. This situation arises in part from their requirement for a low E_m to initiate growth; it is discussed further in Chapter 3.

2
Classification

The two longest established genera of sulphate-reducing bacteria are *Desulfovibrio* and *Desulfotomaculum*,[1] but in recent years several new genera have been identified (*Desulfobacter, Desulfonema, Desulfobulbus* etc., see below). Apart from their ability to reduce sulphate, they seem to be biologically quite unrelated to each other, except that the genus *Desulfomonas* is very like *Desulfovibrio*. Most affinities they may have to other groups of bacteria have become obscured in the course of evolution. The genus *Desulfovibrio* is the best known, largely because its members are usually relatively easier to isolate and purify; they are usually mesophilic and can be halophilic; they do not form spores. Earlier synonyms for this genus were *Spirillum, Microspira, Vibrio* and *Sporovibrio*;[2] the type species is *Desulfovibrio desulfuricans*. The genus now known as *Desulfotomaculum* can be mesophilic or thermophilic but naturally-occurring helophilic strains are rare. The common thermophilic species. *Desulfotomaculum nigrificans*, was earlier known as *Clostridium nigrificans*. All members of *Desulfotomaculum* form spores. Both genera are Gram-negative. The new genera described by Dr Widdel and Professor Pfennig (Widdel, 1980; Pfennig, Widdel & Trüper, 1981) are distinguished superficially by their morphology, but other characters such as the carbon-substrates they can utilize, their growth habit, motility and so on provide subsidiary distinctions.

Like many other types of bacteria, members of this group can show some degree of phenotypic instability in the sense that lines derived from a defined strain will come to differ from the original, usually in minor ways,

---

[1] Being of Latin derivation, *Desulfovibrio* and *Desulfotomaculum* are etymologically correct spellings; *Desulphovibrio* and *Desulphotomaculum* are incorrect, though they appear in some British journals. Sulphur ought to be spelled 'sulfur', since the Romans had no 'ph', but it is several centuries too late for this error to be corrected (see Kelly, 1982).

[2] Pochon & De Barjac (1954) assigned the name *Sporovibrio ferro-oxidans* to a spore-forming vibrio that oxidized Fe^{2+} anaerobically at the expense of nitrate reduction. The report was brief and the strain was lost; whether it had any biological relationship to the spore-forming sulphate-reducing bacteria is not known

after maintenance in different laboratories. I have experienced this phenomenon with the Hildenborough strain of *D. vulgaris*, whose cultural behaviour had changed slightly after a few years in a laboratory in India and in one in the USA; it also changed more impressively after two years in chemostat culture, to a virtually non-motile phenotype. Anecdotal information of this kind is probably available from many workers in this research area; an instance was documented by Kobayashi & Skyring (1982) in which three lines of the Miyazaki strain of *D. vulgaris*, much studied by Japanese workers, had apparently diverged. In this monograph I have referred as far as possible to strains available from recognized culture collections. These hold them, when possible, as lyophilized forms in which phenotypic variation is presumably minimal.

The taxonomy of the sulphate-reducing bacteria is in an unsatisfactory state, having become confused in the 1920s to 1940s by the prevalence of impure cultures and the use of inappropriate culture media (see Chapter 3). These points matter, because impure cultures are capable of metabolizing a wider range of carbon sources than are pure ones (e.g. Kimata, Kadota & Hata, 1955b) and, with *Desulfovibrio*, the presence or absence of a reducing agent influences considerably the apparent range of carbon sources attacked (Grossman & Postgate, 1953). Baars (1930) regularly added sterile H_2S-water to his cultures; the reducing effect of this may account for the fact that his cultures used a wide range of carbon sources for growth, whereas most subsequent workers have recorded a very narrow range of utilizable carbon sources. A small amount of yeast extract is sometimes added to cultures to enhance growth; when tested in media with yeast extract, a strain of *Desulfovibrio* will sometimes show a widened substrate range.

Pure cultures of *Desulfovibrio* or *Desulfotomaculum* have been available for some decades now, and so have prescriptions for suitable media, but still the taxonomic picture is imperfect. At the root of the problem is the relatively small number of diagnostic properties that one can assign. The primary taxonomic character is dissimilatory sulphate reduction; in a monumental survey of 92 *Desulfovibrio* isolates using 116 biochemical characters, Skyring, Jones & Goodchild (1977) found only 26 subsidiary characters to be of taxonomic value, and several of those were of only limited use. Nomenclature must be based on taxonomy, but if the taxonomy is faulty what can one do? Table 1 is a working classification based on that of Campbell & Postgate (1965, 1969) and Postgate & Campbell (1966), up-dated with newly named genera and species even where these are of uncertain status. It is not seriously inconsistent with the statistical

Table 1. A key to the classification of sulphate-reducing bacteria[a]

	Form	Flagella	Motility	Spores	Desulfoviridin	% G+C[c]	Cytochrome[d]
Desulfovibrio							
desulfuricans	Vibrio	Single, polar	+	−	+[b]	59	c_3
vulgaris	Vibrio	Single, polar	+	−	+	65	c_3
gigas	Spirilloid	Lophotrichous	+	−	+	65	c_3
africanus	Sigmoid	Lophotrichous	+	−	+	65	c_3
salexigens	Fat vibrio	Single, polar	+	−	+	49	c_3
thermophilus	Rod	Single, polar	+	−	−	NR[f]	c_3
baculatus	Rod	Single, polar	+	−	+	57	b,c
baarsii	Vibrio	Single, polar	+	−	+	NR[f]	NR[f]
sapovorans	Vibrio	Single, polar	+	−	−	53	b,c
Desulfobacter							
postgatei	Ellipsoidal rod	−	−	−	−	46	b,c
Desulfobulbus							
propionicus	Lemon/onion shape	−	−	−	−	60	b,c
Desulfococcus							
multivorans	Spheres	−	−	−	+	57	b,c
Desulfonema							
limnicola	Long filament	−	gliding	−	+	34–5	b,c
magnum	Long filament	−	gliding/rolling	−	−	42	b,c
Desulfosarcina							
variabilis	Packages, irregular	t[e]	t[e]	−	−	NR[f]	NR[f]
Desulfotomaculum							
nigrificans	Rod	Petritrichous	'tumbling'	+	−	49	b
orientis	Fat vibrio	Petritrichous	+	+	−	45	b
ruminis	Rod	Petritrichous	+	+	−	49	b
acetoxidans	Rod	Petritrichous	+	+	−	37	b
antarcticum	Rod	Petritrichous	+	+	−	NR[f]	b

[a] More details are available in *Bergey's Manual* (Krieg, 1983), Widdel (1980) and Pfennig, Widdel & Trüper (1981).
[b] One aberrant desulfoviridin – negative strain (Norway 4) exists, see text.
[c] % Guanine+cytosine in DNA, see text.
[d] Principal cytochrome(s).
[e] 't' signifies transient packages of irregular non-motile cells which release motile ellipsoid rods.
[f] Not recorded.

Table 1 (*Cont.*)

Growth with:	Lactate + sulphate	Pyruvate + sulphate	Pyruvate	Formate + sulphate	Acetate + sulphate	Glucose + sulphate	Malate + sulphate	Benzoate + sulphate	Other[g]	Thermo-phily	NaCl needed
Desulfovibrio desulfuricans	+	+	−	±	−	v[m]	+	−	Choline minus sulphate	−	−
vulgaris	+	+	−	±	−	−	−	−		−	−
gigas	+	+	−	±	−	NR[f]	−	−		−	−
africanus	+	+	−	±	−	NR[f]	+	−		−	−
salexigens	+	+	−	±	−	NR[f]	+	−		−	+
thermophilus	+	+	−	±	−	NR[f]	−	NR[f]		+	−
baculatus	+	+	−	±	−	−	+	NR[f]		−	−
baarsii	−	−	−	+	+	−	−	NR[f]	Fatty acids to C_{18}	−	−
sapovorans	+	+	+	−	−	−	−	NR[f]	C_{odd} fatty acids	−	−
Desulfobacter postgatei	(±)[h]	−	−	−	+	−	−	−	Vitamins required[k]	−	+
Desulfobulbus propionicus	+	+	+	−	−	−	−	−	Propionate	−	−
Desulfococcus multivorans	+	+	+	+	+	−	−	+		−	−

Table 1 (*Cont.*)

Growth with:	Lactate + sulphate	Pyruvate + sulphate	Pyruvate	Formate + sulphate	Acetate + sulphate	Glucose + sulphate	Malate + sulphate	Benzoate + sulphate	Other[g]	Thermophily	NaCl needed
Desulfonema											
limicola	+	+	−	+	+	−	−	+		−	+
magnum	−	−	−	±[j]	+	−	+	+		−	+
Desulfosarcina											
variabilis	+	+	+	+	+	−	−	−		−	+
Desulfotomaculum											
nigrificans	+	+	+	−	−	v[m]	NR[f]	NR[f]		+	−[1]
orientis	+	+	−	−	−	−	NR[f]	NR[f]		−	−
ruminis	+	+	+	±[j]	−	−	NR[f]	NR[f]		−	−
acetoxidans	−	−	−	−	+	−	−	−	Vitamins required[k]	−	−
antarcticum	+	NR[f]	NR[f]	−	−	+	−	NR[f]		−	−

[g] Substrates of diagnostic interest.
[h] Slow growth of 2 isolates recorded by Widdel & Pfennig (1981*a*).
[j] Some acetate required for growth with formate.
[k] p-aminobenzoate + biotin for *D. postgatei*; biotin alone for *Dm. acetoxidans*.
[l] Nazina & Rozanova (1978) described a halophilic species, see text.
[m] 'v' signifies variable: some strains +, some −.
[n] Badziong, Thauer & Zeikus (1978) described two strains, *D. vulgaris* according to DNA composition, which grew in sulphate-free pyruvate medium.

analysis of Skyring, Jones & Goodchild (1977) and it closely resembles the classification used by Pfennig, Widdel & Trüper (1981), but future taxonomic studies will no doubt impose subdivisions and deletions on the scheme given in Table 1. Not all the types listed there are accepted in *Bergey's manual of systematic bacteriology* (Krieg, 1983); new editions should be consulted to up-date the information in Table 1.

Several features of Table 1 require amplification and an appropriate commentary follows.

Morphology

Desulfovibrio. Morphologically most desulfovibrios are curved or sigmoid (Fig. 2) but departures from the rule exist. For example, the Berre strains and the rather unusual strain Norway 4 of *Desulfovibrio desulfuricans* are usually straight, as is the species *D. baculatus* (an incompletely identified strain is shown in Fig. 2). Desulfovibrios are prone to pleomorphism in old cultures or in incompletely satisfactory environments: they tend to form spirilloid forms. *D. gigas* is unusual because of its large size, spirilloid appearance and intracellular zones of low contrast (see Fig. 3). For many years it was thought to be a unique isolate from Etang de Berre in France but its re-isolation from sewage has been reported (Schoberth, 1973).

Morphology, in fact, is not a good guide to the identification of sulphate-reducing bacteria because, though earlier workers may have obtained an exaggerated idea of their pleomorphism as a result of unwittingly studying impure cultures, it is nonetheless true that even a single pure strain of *Desulfovibrio* may form vibrio, spirilloid, semilunar, straight and sometimes half-empty coccoid forms in response to age and environment. Exceptionally long forms of *Desulfovibrio* are, in my experience, common in response to stresses such as inappropriate salt concentration, magnesium deficiency, a low temperature or a sub-inhibitory concentration of an antibacterial substance, but they also appear in normal cultures; an example is shown in Fig. 3. The type strain of *Desulfovibrio salexigens* strain British Guiana readily assumes a coccoid form. *D. nigrificans* tends to show filamentous growth when cultured near its lower temperature limit.

For morphological comparison it is advisable to use standardized growth conditions. A young culture should be examined (three to six days for mesophiles) after culture in medium B (Table 2, p. 32) or its nearest practicable equivalent. Slow-growing genera may require much longer incubation. The NaCl concentration should match that of the habitat from which the organism was obtained, otherwise pleomorphs will be

Classification

Fig. 2. Phase contrast photomicrographs of diverse sulphate-reducing bacteria. 1, *Desulfovibrio desulfuricans*, strain Essex 6. 2, *D. vulgaris*, strain Hildenborough. 3, *D. baarsii*. 4, *D. sapovorans*. 5, *D. gigas*. 6, *D. africanus*, strain Benghazi (continued overleaf).

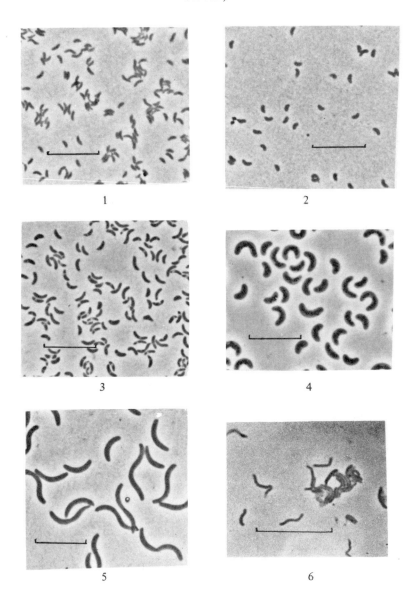

Fig. 2. *(continued)*
7, *D. baculatus* type. 8, *Desulfococcus multivorans*. 9, *Desulfobacter postgatei*. 10, *Desulfobulbus propionicus*. 11, *Desulfosarcina variabilis*. 12, *Desulfonema limicola*. Bar markers 10μm. Photomicrographs by courtesy of Dr F. Widdel except for *Desulfovibrio africanus*, kindly taken by Dr C. Dow.

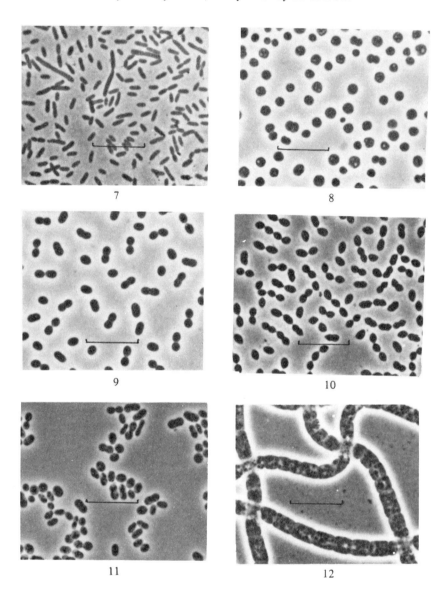

Classification

Fig. 3 Phase contrast photomicrographs of sulphate-reducing bacteria. 1, *Desulfotomaculum nigrificans*, a lenticulate form is arrowed. 2, *Desulfotomaculum acetoxidans*, a: cells from an agar colony showing spores and areas reminiscent of gas vacuoles, b: vegetative cells from a liquid medium (photomicrographs by courtesy of Dr F. Widdel and Professor N. Pfennig.). 3, Stressed form of *Desulfovibrio gigas* which has become a 'snake'; two normal forms are also visible. Bar markers 10μm.

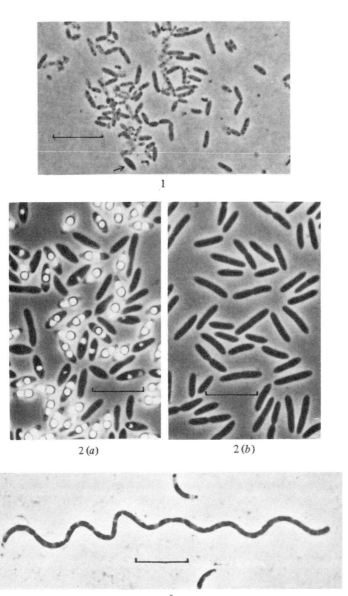

present. A virtue of medium B is that the excess of iron salt traps much of the sulphide formed by the organisms; high sulphide concentrations favour coccoid pleomorphs of *Desulfovibrio*. Yet even when culture conditions are standardized, morphology may vary widely among strains. The species *Desulfovibrio desulfuricans* includes the straight Berre strains and *D. vulgaris* includes the remarkable *croissant*-shaped Groningen strain (see Appendix 1), although most representatives of these species are typical small vibrios.

Figs. 2 and 3 provide a collection of photomicrographs of various types of sulphate-reducing bacteria from 'normal' laboratory cultures; others are available in Pfennig, Widdel & Trüper (1981).

Flagellar morphology is a fairly good taxonomic criterion, though cells lose flagella easily if handled roughly (bacteria from vigorously stirred cultures are often deflagellated and non-motile). Most desulfovibrios are monotrichous (Fig. 4). However, completely non-flagellate strains of *Desulfovibrio desulfuricans* are known, the archetype being the normally (see p. 21) non-motile strain 'Teddington R' (NCIB No. 8312). Some species of *Desulfovibrio* typically have lophotrichous or double flagella.

The genus *Desulfovibrio* is divided, in Table 1, into nine species. The original species was *D. desulfuricans* and the other species have become separated from that group by systematists because they seemed to have relatively stable and homogeneous distinctive characters. It follows that the *desulfuricans* group of *Desulfovibrio* is the least homogeneous: it is more likely than other species to include taxonomically distinct types which are not yet recognized by systematists. The genus *Desulfovibrio* as a whole still lacks sufficient taxonomic criteria and stable distinctive taxonomic characters for a satisfactory classification to be set up.

Desulfotomaculum. In contrast to desulfovibrios, desulfotomacula are usually straight. (Note, however, that *Desulfotomaculum orientis* was classified as a desulfovibrio when first isolated because of its curved appearance: Adams & Postgate, 1959.) Lenticulate (swollen lens-shaped) forms sometimes occur among the thermophilic *D. nigrificans* and may represent aborted sporulating forms (see Fig. 3). Stress, or cultivation at low temperatures, induces long, filamentous pleomorphs which can sometimes form tangled skeins of microbes.

The flagella of desulfotomacula are usually peritrichous (Fig. 4) and often appear to originate laterally. However, as with *Desulfovibrio*, the flagella are easily lost during manipulation and seemingly monotrichous forms can sometimes be observed which are probably artefacts.

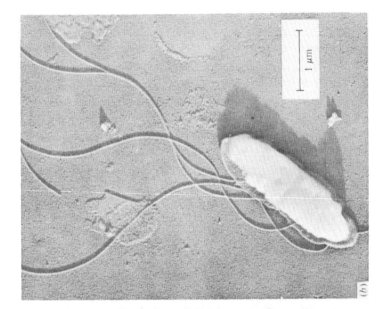
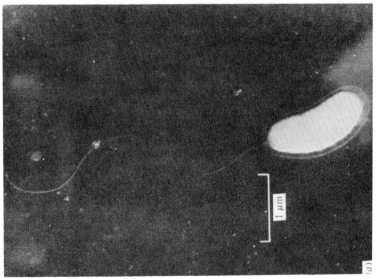

Fig. 4. Electron micrographs of sulphate-reducing bacteria. (a) *Desulfuricans* strain El Agheila Z. (b) *Desulfotomaculum nigrificans* (courtesy of Professor L. L. Campbell).

Spore formation provides the absolute distinction between *Desulfotomaculum* and the other sulphate-reducing bacteria, yet sporulation by *Desulfotomaculum* is not a predictable event: a strain may be subcultured many times without forming a visible spore and may then, for no apparent reason, form up to 90% spores. Donelly & Busta (1980) advised mushroom compost as a medium for maximum sporulation; Widdel & Pfennig (1981*b*) observed that *Desulfotomaculum acetoxidans* sporulates readily when grown with acetate but never with butyrate. Its spores include vacuoles apparently containing gas (Fig. 3). However, even where no spores are visible under the microscope, some can usually be detected by their heat resistance. If 1 ml of culture, after heating at 80 °C for five minutes (allow time to warm up), grows on dilution into medium B, this is presumptive evidence for sporulation. However, spores of *D. nigrificans* can survive for at least 30 min at the boiling point of water.

D. acetoxidans is often found in faeces and intestinal contents; it is probably an inhabitant of the intestinal tract.

Desulfobacter. This genus is represented by only one species, *D. postgatei* (Fig. 2), and consists of stubby rods whose sizes differ according to strain (Widdel & Pfennig, 1981*a*).

Desulfococcus. This genus is also represented by one species, *D. multivorans* (Fig. 2); it appears under phase contrast as spheres, often in clusters evenly separated by capsular material.

Desulfobulbus. The representative species, *D. propionicus*, has a remarkable lemon- or onion-shaped morphology: tapered spheres (Fig. 2; Widdel & Pfennig, 1982).

Desulfosarcina. The one genus *D. variabilis* displays packages of irregular cells which can release small rod-shaped or ellipsoidal individuals (Fig. 2).

Desulfonema. Long filaments capable of gliding and rolling motion: Widdel (1980) described *D. limicola* as 'a few' mm long (Fig. 2) and the giant *D. magnum* as 'several' mm long. *D. limicola* leaves an empty sheath as it migrates. Desulfonemas are tactophilic: growth is restricted to solid surfaces, sometimes as 'spider web-like layers' (Widdel, 1980).

Gram reaction

All sulphate-reducing bacteria are Gram-negative except *Desulfonema* species, which strain irregularly but Gram-positively. Very young cultures of *Desulfotomaculum nigrificans* sometimes show a proportion of Gram-positive cells.

Motility

This property can provide a useful preliminary guide to identifying an unknown culture, but it is not definitive. For example, strains of *Desulfovibrio* capable of motility will sometimes appear non-motile in cultures of high sulphide concentration or after prolonged mechanical agitation (e.g. in a stirred chemostat). Non-motile strains of *Desulfovibrio* have been described, such as strain Teddington R (NCIMB 8312, see appendix 1) but Miller & Wakerley (1966) reported that it was motile in their hands. Conditions determining motility in this group require systematic investigation. Motile desulfovibrios generally show rapid progressive motility under the microscope, contrasting with desulfotomacula which show relatively non-progressive 'tumbling and twisting' motility.

Desulfobacter, *Desulfobulbus* and the free forms of *Desulfosarcina* can sometimes be motile, but isolates differ. *Desulfococci* are non-motile. Desulfonemas show a 'gliding movement often with simultaneous rotation' (Widdel, 1980) often leaving tracks as channels or tubes in sediments.

Cytochromes and chemical composition

The principal cytochromes of *Desulfovibrio* species are called cytochromes c_3 and have a visible absorption peak at 552–553 nm which is characteristic of the species. However, other cytochromes are also present (see Chapter 5). Cytochromes of the *b* and *c* type are present in most of the new genera of sulphate reducer; presence of a cytochrome *b* on its own seems to be diagnostic for *Desulfotomaculum* species. Cytochromes c_3 show electrophoretic and compositional differences (Drucker & Campbell, 1969; Drucker, Trousil & Campbell, 1970; Singleton, Campbell & Hawkridge, 1979) which may be of diagnostic value. Boon *et al.* (1977) suggested that differences among long-chain mono-enoic acids, present in these bacteria, could be used for classification; Ueki & Suto (1979) and Ueki, Azuma & Suto (1981) regarded the cellular fatty acid content as a diagnostic feature. Skyring, Jones & Goodchild (1977) considered that the electrophoretic mobilities of the ATP sulfurylases, sulphite reductases and APS reductases (see Chapter 5) of various strains had diagnostic value.

The desulfoviridin test

This is a simple test for certain *Desulfovibrio* species, particularly the ones most commonly isolated, which is very reliable when a positive result is obtained on a population of vibrios (Postgate, 1959*b*). It can be used on crude enrichment cultures without isolating the organism; about 15 ml of culture is centrifuged down, the damp pellet is resuspended in residual medium and one drop of 2-M NaOH is added under ultraviolet light at 365 mμ. A red fluorescence due to the release of the chromophore of the pigment desulfoviridin (see Chapter 5) is evidence for the presence of several species of *Desulfovibrio* (for the precise species, see Table 1). The desulfoviridin test is also shown by *Desulfonema limicola* and *Desulfosarcina variabilis*, but their morphologies are highly distinctive and they are unlikely to be confused with *Desulfovibrio*.

Strains of *Desulfovibrio desulfuricans* have been described which do not show a positive test, for example the Norway 4 strain of *D. desulfuricans* (Miller & Saleh, 1964). Rozanova & Nazina (1976) described the new species *D. baculatus* (Table 1), which is rod-shaped and desulfoviridin-negative. It may not be coincidence, therefore, that the Norway 4 strain is also rod-shaped, unlike most other strains of *D. desulfuricans* except the Berre strains. Perhaps Norway 4 should be reclassified as *D. baculatus*. The Berre strains should not, since they are desulfoviridin-positive. *Desulfovibrio baarsii* and *D. sapovorans* are also desulfoviridin-negative.

DNA composition

The DNA base ratios (% guanine + cytosine) of the various species are numerically close together, though the precise values depend on the manner of calculation (Skyring & Jones, 1972, also p. 8). The numbers for *Desulfovibrio* in Table 1 are mainly those corrected by Dr Skyring and his colleagues, rounded off to the nearest whole number. The similarity between the values for *Desulfotomaculum*, *Desulfobacter* and *Desulfovibrio salexigens* is probably a coincidence implying no real genetic overlap. Other possible overlaps (e.g. *Desulfobulbus* and the 59% G + C cluster of *Desulfovibrio*; *D. baarsii* and *Desulfococcus* at 57%) are also probably coincidences. The genus *Desulfovibrio* includes three tight clusters of DNA composition (~ 49, ~ 59 and ~ 65) which is unusual: most bacterial genera show a spread of DNA base ratios.

Utilization of carbon sources

The limited substrate specificities of sulphate-reducing bacteria are discussed further in Chapter 5. For Table 1 I have chosen those substrates and conditions which appear to have some diagnostic value. In the genera *Desulfovibrio* and *Desulfotomaculum* growth with acetate in place of lactate is rare but no longer unknown (see Chapter 1, pp. 5–6); acetate is the normal end-product of growth on lactate and acetate-utilizers are separated into distinct species (*D. baarsii* or *Dm. acetoxidans*). Acetate can, of course, be assimilated by several *Desulfovibrio* species in the mixotrophic reaction involving H_2, CO_2 and acetate (pp. 7–8). Growth with sugars is rare and seemingly restricted to some strains of thermophilic desulfotomacula; *Desulfotomaculum antarcticum* is at the time of writing the only sugar-utilizing, mesophilic sulphate reducer (Iizuka, Okahazi & Seto, 1969). Use of benzoate and long-chain fatty acids may be diagnostic for some of the new 'omnivorous' sulphate-reducing bacteria such as *Desulfococcus*, *Desulfosarcina* and *Desulfonema* species. Utilization of propionate but not acetate, so that acetate is an end-product, is the diagnostic metabolic process of *Desulfobulbus propionicus* (Widdel, 1980).

Salt relations

This character includes only absolute requirements for NaCl above about 0.7%. Many desulfovibrios are tolerant of up to 6 or 10% salt, see Chapter 3.

Thermophily and temperature relations

Mesophilic desulfovibrios grow best at about 30° and have an upper temperature limit between 45 and 48 °C. Professor L. L. Campbell and I have independently trained them to tolerate 50 °C but they are then very unstable and killed by an increment of 0.5 °C. Some desulfotomacula have an optimum temperature for growth around 37° (e.g. *D. ruminis*, *D. acetoxidans*) consistent with an intestinal habitat. Thermophilic desulfotomacula have optima around 60 °C and lower limits at about 35 °C, but they acclimatize very slowly to 30 °C. For diagnostic differentiation, a growth test at 55 to 60 °C is adequate but at least one subculture at that temperature must be demonstrated (see Chapter 1, pp. 6–7). The strain of *D. thermophilus* reported by Rozanova & Khudyakova (1974) was the

Fig. 5. Ouchterlony plate of *Desulfovibrio* species. Plate illustrates the immunological cross-reactions between species of *Desulfovibrio* and the absence of such reactions with *Desulfotomaculum*. Anti-serum to whole cells of *Desulfovibrio vulgaris* strain Hildenborough (NCIMB No. 8303) was raised in rabbits. About 0.2 ml was placed in the centre wells (hatched in sketch) made in buffer set with agar in a Petri dish. Suspensions of species indicated with NCIMB numbers, equivalent to 50–100 mg dry wt, were placed in the peripheral wells. Lysis of the cells occurred followed by diffusion of cell constituents; where these met the homologous component of the anti-serum a precipitation band was formed at a characteristic distance from both wells. The bands were photographed in conditions which enhanced light scattering after two days. The sketch is a guide to the precipitation lines seen.

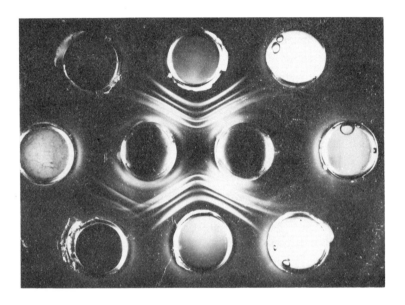

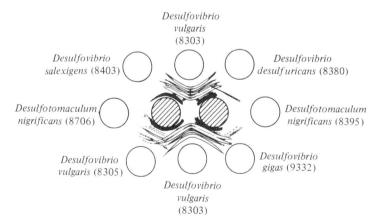

Notice excellent cross-reaction (> 7 bands) between anti-serum and its homologous 8303 cells and almost equivalent (> 6 bands) reaction with another strain (NCIMB No. 8305). Fair reactions with *Desulfovibrio gigas* (NCIMB No. 9332, 2 bands), *Desulfovibrio salexigens* (NCIMB No. 8403, 1 band) and *Desulfovibrio desulfuricans* (NCIMB No. 8380, 1 band). No cross-reaction with the two strains of *Desulfotomaculum nigrificans*.

---

first example of a thermophilic desulfovibrio to be described; a new thermophile (see p. 29) may be a relative.

That completes the commentary on Table 1, but more must be said on other possible taxonomic criteria, and on the nomenclature of sulphate-reducing bacteria.

'Hibitane' resistance

Sulphate-reducing bacteria, particularly desulfovibrios, can show exceptional resistance to inhibitors (see Appendix 2 and Saleh, Macpherson & Miller, 1964). Resistance to 'hibitane' (bis-p-chlorophenyldiguanidohexane diacetate) and certain other microbicides once showed considerable promise as a diagnostic character but, as more strains were examined, it proved less definitive. Nevertheless, in conjunction with other properties hibitane resistance seems to retain some value within the genus *Desulfovibrio* (Skyring et al., 1977). *D. salexigens* strains are generally highly resistant, growing with 100 to 1000 mg/ml; *D. desulfuricans* strains tolerate 10–25 mg/ml; the *vulgaris–africanus–gigas* group tolerate only about 2.5 mg/l. Desulfotomacula are especially sensitive; *D. nigrificans*, *D. orientis* and *D. ruminis* generally tolerate only 0.25 to 1 mg/ml. Data for other species in Table 1 are not available.

DNA–rRNA homology

Ribosomal RNA (rRNA) is highly conserved and homology between DNA from one organism and rRNA from another is a measure of relatedness. Pace & Campbell (1971) studied the homology of various rRNAs to DNA from *Desulfovibrio vulgaris* and obtained the following order (% homology in parentheses). *Desulfovibrio vulgaris* = *Desulfovibrio desulfuricans* (91%) > *Desulfovibrio salexigens* (67%) > *Desulfovibrio africanus* (59%) > *Desulfotomaculum nigrificans* (39 to 49%).

26 Classification

Fig. 6. Ouchterlony plate of *Desulfotomaculum* species. Plate illustrates identification of *D. ruminis* as a relative of *D. nigrificans*. Immunodiffusion set up as for Fig. 5 and photographed after two days. Centre well (hatched in sketch) contained rabbit anti-serum to whole cells of *D. nigrificans* strain Delft 74T (NCIMB No. 8359). Outer wells contained suspensions of 50–100 mg dry wt of the strains indicated by name and NCIMB number (see Appendix 1). The sketch is a guide to precipitation lines seen.

Notice strong (> 4 line) reaction between anti-serum and its homologous 8395

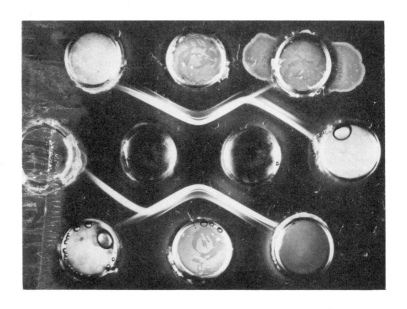

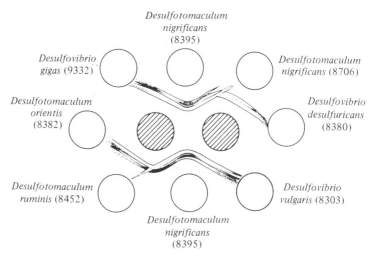

cells, good (3 line) cross-reaction with another strain (NCIMB No. 8706), clear cross-reaction with *D. ruminis* (NCIMB No. 8452) and absence of reaction with the three species of *Desulfovibrio*. A reaction with *Desulfotomaculum orientis* (NCIMB No. 8382) was expected but was not shown with this anti-serum; a comparable test with anti-serum to *Desulfotomaculum ruminis* (NCIMB No. 8542) later identified *Desulfotomaculum orientis* as a taxonomic relative.

Serology

Anti-sera prepared towards whole cells of members of a given species show pronounced cross-reactions with other strains of the species and, usually, some cross-reactions with other members of the genus (Figs. 5 and 6). Positive crosses are thus valuable indications of taxonomic similarities but negative serological cross-reactions cannot, alone, be accepted as indicating specific or generic differences. Serology played an important part historically in identifying '*Clostridium nigrificans*' as *Desulfotomaculum nigrificans* and in clarifying the taxonomy of the desulfotomacula (Campbell & Postgate, 1965; see also Figs. 5, 6). However, the quality of the anti-serum for diagnostic purposes is probably affected by the choice of type strain against which the antibody is raised, as well as the responsiveness of the animal in which it is raised. For example, Sefer & Pozsgi (1968) found serological heterogeneity among isolates of *Desulfovibrio*, and Abdollahi & Nedwell (1980) found remarkably little cross-reactivity even within species of *Desulfovibrio* with their anti-sera. Smith (1982) also found only limited cross-reactivity using immunofluorescence.

The salt-requiring subspecies aestuarii

Salt-tolerant and marine strains of *Desulfovibrio* are common. If they have an absolute requirement for more than 1% NaCl and yet do not belong to the species *Desulfovibrio salexigens*, they are given subspecies status (earlier they were called 'varieties'): *D. desulfuricans aestuarii* and *D. vulgaris aestuarii* both exist. Comparable salt-requiring strains of *Desulfotomaculum* are rarely found, but Nazina & Rozanova (1978) isolated such an organism, a thermophile requiring at least 1% NaCl, from oil strata, and termed it *D. nigrificans* subspecies *salinum*.

The variety azotovorans *of* Desulfovibrio

The property of nitrogen fixation was once thought to be restricted to two strains of *Desulfovibrio desulfuricans* (the Berre strains) which were assigned varietal status as *D. desulfuricans azotovorans* (Postgate & Campbell, 1966). With the realization that many strains and species fix nitrogen (Riederer-Henderson & Wilson, 1970; Postgate, 1970a) the variety or subspecies *azotovorans* must lapse.

'Desulfovibrio hildenborough'

Many of the early workers on sulphate reduction, lacking any satisfactory taxonomic base, named their isolates after their place of origin. Numerous strain descriptions are collected in Appendix 1; examples are:

Desulfotomaculum nigrificans	Delft 74T	Isolated by R. L. Starkey at Delft in Holland
Desulfovibrio vulgaris	Hildenborough	By M. E. Adams, in soil from Hildenborough, England
Desulfovibrio vulgaris	Wandle	By M. E. Adams, from the River Wandle, England
Desulfovibrio desulfuricans	El Agheila Z	By M. E. Adams, in mud from El Agheila, Libya
Desulfovibrio desulfuricans	Berre E	By J. Le Gall, in water from Etang de Berre, France
Desulfonema limicola	Jadebusen 5ac10	By F. Widdel, from brackish water near Jadebusen (North Sea)
Desulfovibrio africanus	Benghazi	By M. E. Adams, in water from Benghazi, Libya
Desulfovibrio salexigens	British Guiana	By M. E. Adams, in marine mud from the then British Guiana

This strain designation is now becoming replaced by National Collection of Industrial and Marine Bacteria numbers. Not before time, because *Desulfovibrio hildenborough*, a misnomer resulting from a confusion of specific and strain names, has made its appearance at least twice in the published literature (Thomas, 1972; Spiro, 1977).

Desulforistella, Desulfomonas pigra *and the rod-shaped desulfovibrios*

Many rod-shaped sulphate-reducing bacteria have been reported, including *Desulforistella* (Hvid-Hansen, 1951; no longer a valid name according to Dr N. Krieg, personal communication), the Berre strains of *Desulfovi-*

brio desulfuricans (Le Gall, Senez & Pichinoty, 1959), a fluorescent species resembling *Desulfovibrio vulgaris* reported by Jones (1971*b*), *Desulfovibrio baculatus* (Rozanova & Nazina, 1976) and *Desulfomonas pigra* (Moore, Johnson & Holderman, 1976). Rozanova & Nazina (1976) suggested that a new genus, rather than a new species, might be necessary to accommodate such organisms. *Desulfomonas pigra* was examined and described by techniques (e.g. broth cultures) not normally used with sulphate-reducing bacteria; insofar as comparison can be made, the organism is a relatively large non-flagellate rod with *b*- and *c*-type cytochromes, desulfoviridin-positive (Sperry & Wilkins, 1977) and able to utilize conventional substrates for *Desulfovibrio*. Its principal distinctive feature is the high guanine and cytosine content of 66.67%. I have chosen not to include it in Table 1 because the question of its relation to the Berre strains, strain Norway 4 and *Desulfovibrio baculatus* requires careful investigation. Pfennig, Widdel & Trüper (1981) accepted it as a distinct genus and species. I have also chosen to exclude a thermophilic acetogenic rod, of low G+C ratio, possessing cytochrome c_3 but not desulfoviridin, designated *Thermodesulfobacterium commune* by Zeikus *et al.* (1983). It seems to bear a relationship to *Desulfovibrio thermophilus* similar to that of the Norway 4 strain to *D. desulfuricans*. Therefore an exhaustive comparison with *D. thermophilus*, which has taxonomic priority, would seem to be necessary before it is accepted as an entirely new genus and species.

Taxonomic problems can become ends in themselves. In this chapter I have dismissed earlier problems and some which seem rather minor – they may be traced from the reviews cited in Chapter 1. It must be obvious that the key provided in Table 1 is far from immutable; the new organisms described by Dr Widdel and Professor Pfennig may catalyse the discovery of more types; the surveys by Skyring, Jones & Goodchild (1977), Sefer & Pozsgi (1968) and Ueki, Azuma & Suto (1981) suggest that some further subdivision and/or reorganization of *Desulfovibrio* will become necessary. Conflicts continue to be reported between cultural or other taxonomic data obtained in different laboratories with the same species and even with putatively identical strains; a rational taxonomy of this group is still some way off. But Table 1 covers most of the sulphate-reducing bacteria now available in research laboratories and culture collections as well as most of those commonly encountered in the natural environment. Thus it represents a working classification which can usefully be adopted until sufficient new strains have been examined and compared and sufficient definitive characters have been established for a better classification to be proposed.

3

Cultivation and growth

There is an element of art, as well as science, in cultivating the sulphate-reducing bacteria in pure culture, but one of the major prerequisites is simple: the redox potential (E_h) of the environment must start around −100 mV. This means that mere exclusion of air is not sufficient to ensure growth (a boiled-out lactate + sulphate medium would have an E_h of about +200 mV under N_2; with 5-mM Na_2S the value would be about −220 mV). The presence of a redox-poising agent is necessary.[1] Large inocula produce such an agent – H_2S – by their resting metabolism; small inocula require added H_2S (usually as sterile Na_2S). A thiol compound such as cysteine or sodium thioglycollate may be used with desulfovibrios or desulfotomacula but thioglycollate is not recommended for other genera by Pfennig, Widdel & Trüper (1981). These authors advised Na_2S or sodium dithionite (Widdel & Pfennig, 1977) as reductants. Once a vigorously growing culture is established, the reductant can often be omitted provided an inoculum is used which is large enough to carry over adequate Na_2S. Media are incubated in the absence of air using the procedures customary in handling anaerobes: completely filled and stoppered vessels, vessels with plugs containing alkaline pyrogallol to remove the last traces of oxygen, or vessels in containers filled with an inert gas such as N_2. For diagnostic purposes, media are often prescribed which contain about 0.5% of a ferrous salt: this forms a black precipitate of FeS when sulphide is formed, so blackening of the medium as a whole, or the zone round a colony, is evidence for bacterial sulphate reduction. A detailed review of practical methods for enriching, isolating, checking the purity of and counting sulphate-reducing bacteria is beyond the scope of this mon-

---

[1]It is a poor reflection upon the efficiency of communication among scientists that, though the need for a low E_h has been known since around 1953, and has been publicized repeatedly, I attended a meeting of Canadian microbiologists in 1977 at which a senior scientist still presented data he believed to be quantitative yet obtained with unpoised media.

ograph. Postgate (1965*b*) gave details of experimental procedures and Pankhurst (1971) presented a variety of media. Pfennig, Widdel & Trüper (1981) described media for non-sporulating 'new' sulphate-reducing bacteria in considerable detail. Every laboratory has its 'pet' media and it is not always clear which details are crucial. Some routine procedures which are generally successful will be presented here.

Table 2 gives recipes for a variety of media used when working on sulphate-reducing bacteria. A commentary on their composition and use follows.

Medium B is a general purpose medium for detecting and culturing *Desulfovibrio* and *Desulfotomaculum* although it is not useful for *Dm. acetoxidans*. Most of the ingredients can be prepared and held as a stock, but the thioglycollate and ascorbate, which may be omitted if the inoculum is a fresh, flourishing culture, should be added, and the pH adjusted, just before autoclaving. The medium should then be used as soon as it is cool because the reductants deteriorate in air at neutral pH values, a process accompanied by a transient purple colour. The precipitate in medium B aids growth of tactophilic strains. It is recommended for long-term storage of strains.

Medium C is a clear medium for mass culture of *Desulfovibrio* and most desulfotomacula, for chemostat culture and for research in which a turbid medium is not desirable. The citrate serves to hold iron and possibly other trace elements in solution and recommended concentrations have ranged from 0.3 to 5 g/l. The latter concentration markedly delays growth of *D. gigas* but is tolerated by *D. vulgaris* and several other desulfovibrios. The concentration quoted (which differs from that in the first edition of this monograph) is satisfactory with all strains handled by the writer so far.

Medium D is a diagnostic medium, testing for sulphate-free growth of desulfovibrios or desulfotomacula (see Table 1).

Medium E is for counting populations of such organisms as black colonies in deep agar, and for isolation of pure cultures. As with medium B, the reductants should be added just before autoclaving, the pH re-adjusted, and the medium used as soon as it cools.

Medium F is based on iron sulphite agar, widely used in food microbiology to diagnose *Desulfotomaculum* (*Clostridium*) *nigrificans*, supplemented with lactate and a magnesium salt to make it suitable for desulfovibrios and some other desulfotomacula as well (Mara & Williams, 1970).

Medium G is a general formulation based on the prescriptions given by Pfennig, Widdel & Trüper (1981) for *Desulfotomaculum acetoxidans*, *Desulfovibrio baarsii* and *D. sapovorans*, *Desulfobacter*, *Desulfonema*,

Table 2. *Some media for sulphate-reducing bacteria (concentrations given in g/litre; for commentary see text)*

Medium B

KH_2PO_4	0.5
NH_4Cl	1
$CaSO_4$	1
$MgSO_4.7H_2O$	2
Sodium lactate	3.5
Yeast extract	1
Ascorbic acid	0.1
Thioglycollic acid	0.1
$FeSO_4.7H_2O$	0.5

Tap water 1 litre, adjust reaction to between pH 7 and 7.5. This medium always contains a precipitate. NaCl should be added for marine strains, or sea-water used in place of tap water.

Medium C

KH_2PO_4	0.5
NH_4Cl	1
Na_2SO_4	4.5
$CaCl_2.6H_2O$	0.06
$MgSO_4.7H_2O$	0.06
Sodium lactate	6
Yeast extract	1
$FeSO_4.7H_2O$	0.004
Sodium citrate.$2H_2O$	0.3

Distilled water 1 litre, pH 7.5 0.2. This medium may be cloudy after autoclaving but should clear on cooling. Add extra NaCl for salt-water strains.

Medium D

KH_2PO_4	0.5
NH_4Cl	1
$CaCl_2.2H_2O$	0.1
$MgCl_2.6H_2O$	1.6
Yeast Extract	1
$FeSO_4.7H_2O$	0.004
Sodium pyruvate	3.5
or choline chloride	1

Distilled water 1 litre, pH 7.5 ± 0.2. Sterilize by filtration; add extra NaCl for salt-water strains. Malate or fumarate have been used as carbon sources for research purposes.

Medium E

KH_2PO_4	0.5
NH_4Cl	1
Na_2SO_4	1
$CaCl_2.6H_2O$	1
$MgCl_2.7H_2O$	2
Sodium lactate	3.5
Yeast extract	1
Ascorbic acid	0.1
Thioglycollic acid	0.1
$FeSO_4.7H_2O$	0.5
Agar	15

Tap water 1 litre. Adjust to pH 7.6 with NaOH after boiling to dissolve agar. Autoclave and use before it solidifies. Add extra NaCl for salt-water strains.

Table 2 (cont.)

Medium F		
Tryptone	10	This recipe is essentially 'Iron-sulphite
Sodium sulphite	0.5	agar' as marketed complete by Oxoid Ltd
Iron citrate	0.5	with lactate and Mg^{2+} supplements.
Agar	12	Prepare according to Oxoid's instructions
Sodium lactate	3.5	but add supplements indicated, pH 7.1.
$MgSO_4.7H_2O$	0.2	

Medium G		
KH_2PO_4	0.2	Distilled water 970 ml. Sterilize by
NH_4Cl	0.3	autoclaving, components marked * below
Na_2SO_4	3	added aseptically later; pH to 7.2 with 2M-
$CaCl_2.2H_2O$	0.15	HCl. $Na_2S_2O_4$ sometimes added as well.
$MgCl_2.6H_2O$	0.4	NaCl to 20 and $MgCl_2.6H_2O$ to 3 for
KCl	0.3	marine strains.
NaCl	1.2	

Additions to medium G:
Selenite, 3 µg*. From autoclaved stock of Na_2SeO_3 3 mg + NaOH 0.5 g/l.
Trace elements, 1 ml*. From autoclaved stock of $FeCl_2.4H_2O$, 1.5 g; H_3BO_3, 60 mg; $MnCl_2.4H_2O$, 100 mg; $CoCl_2.6H_2O$, 120 mg; $ZnCl_2$, 70 mg; $NiCl_2.6H_2O$, 25 mg; $CuCl_2.2H_2O$, 15 mg; $NaMoO_4.2H_2O$, 25 mg/l.
$NaHCO_3$, 2.55 g*. 30 ml of 8.5% w/v solution, filter-sterilized after saturation with CO_2.
$Na_2S.9H_2O$, 0.36 g*. 3 ml of 12% w/v solution autoclaved under N_2.
Vitamins, 0.1 ml*. From filter-sterilized stock of biotin, 1 mg; p-aminobenzoic acid, 5 mg; vitamin B_{12}, 5 mg; thiamine, 10 mg/100 ml.
Growth stimulants, 0.1 ml*. From autoclaved stock of isobutyric acid, valeric acid, 2-methylbutyric acid, 3-methylbutyric acid, 0.5 g of each; caproic acid, 0.2 g; succinic acid, 0.6 g/100 ml, NaOH to pH 9.
Carbon sources*. 1 ml/100 ml final medium of autoclaved stocks including: Na acetate.$3H_2O$, 20%, propionic acid, 7%, n-butyric acid, 8%, benzoic acid, 5%, n-palmitic acid, 5% etc. Free acids neutralized to pH 9; special procedure needed for palmitate, see reference below. Other C-sources can obviously be substituted.
(See Pfennig, Widdel & Trüper, 1981, for fuller details of medium G.)

Desulfobulbus etc. (see Table 1). $Na_2S_2O_4$ may sometimes be used in place of Na_2S (e.g. for *Dm. acetoxidans*: Widdel & Pfennig, 1977). Medium G includes trace elements, vitamins, carbonaceous growth stimulants such as succinate and valerate (these accelerate growth of some rumen anaerobes); the carbon source should be matched to the type of organism being sought or cultivated; the species obtained depends on the precise formulation used. For *Desulfobacter postgatei*, three different concentrations of NaCl and MgCl$_2$ were used and KCl was 0.5 g/l (Widdel & Pfennig, 1977). Trace elements, growth stimulants and vitamins may not always be necessary. Fuller details of the variations of the media listed here as medium G are available in the article by Pfennig, Widdel & Trüper (1981) which should be consulted. Note that *Desulfosarcina* is light-sensitive and must be cultivated in the dark.

Bacteriological procedures for handling sulphate-reducing bacteria

Temperature: Most mesophiles are grown at 30 °C and most thermophiles at 55 °C, but these guidelines are not rigid: for example, Widdel & Pfennig (1977) used 36 ° for *Desulfotomaculum acetoxidans* and 29 ° for *Desulfobacter postgatei* (Widdel & Pfennig, 1981a). Clearly common sense should be used: psychrophiles should be sought at low temperatures and thermophiles at high temperatures, but the likelihood that vegetative (or spore) forms can survive in nature at temperatures inimical to growth should not be neglected.

Sample: Samples of water should be in sterilized vessels, completely filled to exclude air. Samples of soil should be taken from an appropriate level (e.g. that at which an iron pipe is to be laid) and should be transferred immediately to clean, dry, screw-top jars which must be completely filled to exclude air as far as possible. The examination should be carried out as soon as possible, preferably within 24 hours if the 'qualitative' procedure is being used. The samples should be kept cool, but not frozen, until tested.

Qualitative test for most desulfovibrios and desulfotomacula: Place about 2-g and 0.2-g samples (or 2- and 0.2-ml water samples) in sterile stoppered or screw-capped bottles of 10 to 20 ml capacity. Fill up to the brim with fresh medium B. Stopper so that no air is trapped in the bottle and incubate at 30 °C (55 °C for thermophiles). Samples of marine or estuarine origin require approximately 2.5% NaCl in the medium.

The presence of lactate-utilizing sulphate-reducing bacteria is indicated by blackening of the medium (by production of FeS). Fresh aggressive soils blacken in two to three days. Occasionally the 0.2-g sample blackens before the larger one; this is due either to an inhibitor in the soil, or to oxidizing material in the soil which delays growth. Samples taking more than 21 days to blacken, or not blackening at all, either contain few sulphate-reducing bacteria or have suffered excessive exposure to air.

Procedures for the detection and enrichment of sulphate-reducing bacteria capable of utilizing other carbon sources are based on the use of variants of medium G. Blackening does not occur and is not an indicator of growth. Procedures were given in some detail by Pfennig, Widdel & Trüper (1981).

Quantitative test for desulfovibrios: Water samples are diluted in saline as for any conventional bacteriological count. For soils and sediments, a 1-g (± 0.01 g) fresh sample is ground in a sterilized mortar and 9 ml sterile saline (0.8% NaCl normally; 2.5% NaCl for marine soils) is gently 'ground in' to give a fine suspension. This is the first decimal dilution, 1/10 : 1 ml of this solution is added to 9 ml fresh saline to give the second $1/10^2$ dilution, and so on down to about $1/10^6$ (Pochon, 1955). Each dilution, in 1 ml lots, is added to long (15×1 cm) test tubes with sterile bungs, and 6 ml of medium E (held molten at 40 °C) is added. The contents are mixed by inverting the tubes and, once the agar is set, an extra 1 ml of agar is added as a 'plug' to exclude air from the inoculated portion. Incubate at 30 °C (55 °C for thermophiles) in air and count black colonies in the agar. Calculate the original population from the colony number and the dilution. Ideally three to five 'deep agar tubes' should be used for each dilution tested and the number of colonies counted should be 50 to 200.

Occasionally gaseous clostridia accompany sulphate-reducing bacteria and disrupt the agar. Fortunately this happens rarely except with rumen or sewage samples. If it occurs, deep agar counts are not practicable and dilutions should instead be inoculated into quintuplicate 9 ml lots of medium B and incubated anaerobically. Blackened tubes are positives; counts can be estimated from statistical tables of 'Most Probable Numbers' (e.g. Postgate, 1969*a*).

Methods for counting lactate-utilizing sulphate-reducing bacteria on agar plates have been described and were discussed by Pankhurst (1971). In my hands they give results comparable to the procedures just described but two or three days more slowly. For success with plates it is essential to have reliable anaerobic containers for incubation.

The procedures just described for the quantitative estimation of *Desulfovibrio* have been evaluated by taking fresh, pure cultures and comparing the absolute (microscopic) count with the viable count. They agree within the acceptable limits of viable counts of other bacteria. Quantitative enumeration of most *Desulfovibrio* species is thus possible using either plates or deep agar tubes. However, no satisfactory medium has yet been published for quantitative enumeration of *Desulfotomaculum* (see Mara & Williams, 1970) nor are quantitatively proven procedures for organisms such as *Desulfobacter* and *Desulfobulbus*. It is remarkable that, even in the

1980s, 'satisfactory' media for enumerating sulphate-reducing bacteria were being published without checks on their effectiveness with populations of known numerical density; such reports should be viewed with suspicion. Numerical values for desulfotomacula and other non-desulfovibrio genera represent minimum, not absolute, counts of viable organisms.

Isolation of sulphate-reducing bacteria

... the most remarkable fact that even nowadays, 45 years after Beijerinck's pioneer work, the number of laboratories in which pure culture of sulphate-reducing bacteria have been obtained can probably be counted on the fingers of one hand. [Professor A. J. Kluyver, 1940.]

Sulphate-reducing bacteria are present in most soils and waters, but are outnumbered by other types of microbes except in special environments. Hence enrichment of the population with respect to these bacteria is usually necessary before isolation is attempted. Batch culture is usually used for enrichment purposes though the use of chemostats (see p. 40) is rewarding according to Keith, Herbert & Harfoot (1982).

Such enrichment cultures can be prepared from soil, water or other samples using medium B or G just as in the qualitative test procedure described above. Lactate-utilizing sulphate-reducing bacteria may be isolated from the enrichment culture by a simplification of the 'quantitative test' using medium E. The presence of reductants in medium E makes isolation a much less formidable task than it was in the days of Professor Kluyver's remark quoted above. A description of an effective procedure follows.

Distribute approximately 4-ml lots of medium E into six long test tubes (15×1 cm) and keep them at 40 °C. Dilute inoculum from the enrichment by dipping a clean wire, or a closed Pasteur pipette, into the enrichment and then successively into tubes 1 to 6, which is a quick and very successful procedure if the level of agar in the tubes is roughly similar. Set agar and incubate in air at 30 °C (55 °C for thermophiles); in 24–48 hours colonies will appear, the sulphate reducers being black; these should be nicely separated in the fourth or fifth tube by the fourth or fifth day. There is no need to incubate the tubes anaerobically; thermophiles at 55 °C grow more rapidly.

Break a suitable tube at a convenient point and withdraw two or three colonies with a fine Pasteur pipette. Break up each colony in sterile saline (0.5 ml or less), inspect the cell suspension under the microscope. If it appears pure, use suspension as inoculum into medium B. Check a subculture for purity as below. If doubtful, repeat 'Pasteur pipette' dilution as above at once and re-isolate new colonies.

For isolating the non-*Desulfovibrio* genera, and some of the more difficult desulfovibrios and desulfotomacula, Pfennig, Widdel & Trüper (1981) recommended an essentially similar procedure based on medium G, containing $Na_2S_2O_4$ and set with agar, and sealed with wax because prolonged incubation (2–4 weeks) may be needed. They prescribed appropriate variants of medium G for the different types. Since several of these genera are tactophilic or tend to grow in clumps, it is *most important* to disperse such growth in the enrichment culture, or the sample therefrom, before diluting. Neglect of this step may be the reason why workers failed to isolate these bacteria before Widdel (1980). Such isolates must, of course, be checked biochemically for sulphate reduction since blackening, the indicator with medium B, does not occur.

Routine culture

Cultures of most types of *Desulfovibrio* or *Desulfotomaculum*, if well sealed in medium B, will remain viable for a long time (see 'Storage' below). Nevertheless, stocks for routine use should be subcultured at intervals of one to six months, using any of the conventional means for ensuring anaerobiosis (pyrogallol-saturated plugs, Fildes-Mackintosh jars, etc.). Special tubes for sulphate-reducing bacteria were designed by Pankhurst (1967) and are available commercially.

A precaution is necessary regarding the anaerobic atmosphere. Since the reduction of Na_2SO_4, the usual sulphate source, leads to Na_2S formation, and this hydrolyses to $NaOH + H_2S$, sulphate reduction tends to make the medium alkaline and old cultures may die rapidly if the H_2S evaporates. It will evaporate into NaOH in pyrogallol plugs. Use of Na_2CO_3 (15%) + NaOH (5%) as the alkali in the pyrogallol plug will compensate for the effect. Similarly, CO_2 (1%) should be a component of the atmosphere of an anaerobic jar, but too much CO_2 must be avoided or the pH may become too acid before growth starts and thus prevent growth altogether.

If anaerobic jars are used which contain a platinum or palladium catalyst to remove the last traces of oxygen, it is wise to remember that H_2S poisons such catalysts and they may soon become non-functional.

Old cultures tend to grow rather capriciously on subculture, particularly in media such as C or D and, for physiological work, continuous culture in a chemostat is strongly recommended (see p. 40).

Sporulation

Little systematic information is available on media to induce sporulation of desulfotomacula. The matter was discussed in Chapter 2, p. 18.

Purity checks

It cannot be overemphasized that checks for both aerobic and anaerobic contaminants are essential to establish purity.

Contaminant aerobes: plate out in usual manner, using any nutrient agar containing glucose and peptone; no colonies should appear after incubation in air at 30 °C.

Contaminant anaerobes: prepare 25 ml of molten peptone–glucose agar and add sterile ferrous ammonium sulphate to 0.05%. Adjust the pH to 7–7.6 and distribute into five or more 15 × 1 cm sterile plugged test tubes. Cool to 40 °C, make 'Pasteur pipette' dilutions from the test culture, set and incubate at 30 °C (55 °C for thermophiles). If the culture is pure, a relatively small number of exclusively black colonies will appear after three or four days; colonies which are not black are due to contaminant anaerobes. Bubbles, sometimes disrupting the agar, indicate contaminating gasogenic anaerobes.

Storage of stock cultures

Cultures of *Desulfovibrio* or *Desulfotomaculum* in medium B, well sealed against air, are amazingly long-lived at room temperature in a dark cupboard. Cultures so stored in 1937 by the late M. E. Adams were opened in 1955 and subcultures grew the next day. A belief exists among workers in this area that the precipitate of FeS is helpful to survival; though it lacks firm scientific basis, the wise microbiologist will heed it. Such cultures are inconvenient for transport or sending by mail. They may either be set in agar, or freeze-dried cultures may be prepared. A procedure developed for desulfovibrios and desulfotomacula by Mr A. R. MacKenzie of the National Collection of Industrial and Marine Bacteria follows.

Grow eight 20 ml cultures in medium C (or other appropriate medium) in vessels which can be centrifuged. Centrifuge aseptically after growth and pool the organisms in 7 to 10 ml filter-sterilized *mist. desiccans* (1 part nutrient broth, 3 parts serum, glucose to 7.5%). Distribute 5 drop aliquots into about 50 ampoules and freeze-dry centrifugally, followed by secondary drying over P_2O_5. Avoid excessive exposure to air at all stages. Do not add NaCl to suspensions of halophilic strains.

Cultures lyophilized in this way remain viable for at least 10 years; they may be revived in medium B or medium C.

Growth characteristics of sulphate-reducing bacteria

Bacteria for research purposes should be relatively free of FeS precipitate. Medium C contains relatively little iron and is thus suitable for *Desulfovibrio* and *Desulfotomaculum*. These organisms also grow in lactate-containing medium G; for other genera the appropriate carbon sources are needed. Growth of mesophilic sulphate-reducing bacteria is often slow, taking from several days to 2 weeks at 30 °C according to species. Thermophiles grow much faster and, at 55 °C, growth generally takes 12 to 18 hours. One reason for slow growth is that H_2S, the product of respiration, decreases the growth rate and can, at high concentration, slow the growth rate to zero. This phenomenon may be due to the intrinsic toxicity of H_2S to living systems, but it is more likely that it arises because H_2S renders soluble iron insoluble by converting it to FeS. Growth of cultures of *Desulfovibrio* in many media follows a linear rather than exponential course; exponential growth can be obtained in media containing chelating agents to increase the solubility of iron (see Postgate, 1965a) and the citrate in medium C partly serves this function. A medium such as C should yield about 0.4 mg dry wt *Desulfovibrio*/ml at 30 °C in one to two days from a 2% inoculum. Yields of *Desulfotomaculum* are about half of this. Implications of growth yields on ATP economy are mentioned in Chapter 5 (p. 92).

Since growth in batch cultures is frequently non-exponential it is rarely useful to quote doubling times and those found in the literature are often fortuitous (unless the exponential character of growth is clearly stated). However, it is possible to generalize in a qualitative sense: acetate-producing organisms such as the majority of desulfovibrios and desulfotomacula grow rapidly and seem capable of doubling times, at 30 °, as low as 3 to 6 hours. Acetate-utilizing types such as desulfobacters grow much more slowly and, though quantitative data are not available, are probably incapable of doubling times shorter than 20 hours.

Because of the uncertainties introduced by the dependence of growth on E_h, pH and medium composition, and problems caused by accumulation of sulphide, continuous culture rather than batch culture is greatly to be preferred for research purposes. Keith, Herbert & Harfoot (1982) obtained unusual nutritional types of sulphate reducer using a chemostat as an enrichment system. I have run continuous cultures of *Desulfovibrio* for up to two years in medium C and modifications thereof; once established they are less prone to contamination than batch cultures and they supply reproducible inocula for occasions when batch cultures are unavoidable. The most practical type of continuous culture is that known as the chemostat, in which the population density is determined by the concentration of a nutrient (e.g. lactate or sulphate) in the medium and the population grows at a rate lower than its maximum doubling time (μ_{max}). Normal strains of *Desulfovibrio* in conditions permitting exponential growth have $\mu_{max} \sim 3$ hours in medium C and grow conveniently in lactate-limited chemostats at μ between 4 and 7 hours (dilution rates of 0.2 to 0.1). The theory of such chemostats is simple (a clear and sufficient introduction is given by Veldkamp, 1976) and I emphasize that complicated apparatus is quite unnecessary for this group: the vigorous and reliable stirring systems which are needed with aerobes can be dispensed with and for most purposes magnetic stirring is adequate. Given a reliable pump, an effective chemostat system can be assembled from common laboratory glassware, although some purpose-designed components are helpful. Appropriate cheap equipment was described by Baker (1968). However, continuous culture of sulphate-reducing bacteria has four special requirements.

1. The sulphide produced corrodes iron and even, in time, 'stainless' steel. Hence all-glass apparatus such as Baker's (1968) is desirable.

2. Silicone and light rubber tubes are permeable to air, so sufficient O_2 can diffuse inwards to stop growth unless heavy rubber or polyvinylchloride tubing and connections are used.

3. The toxicity of H_2S[1] must not be forgotten when growing and harvesting these bacteria: it is about as toxic as HCN and, though the human nose is ordinarily much more sensitive to it, sensitivity declines rapidly on exposure to the gas. Therefore the continuous culture must include a line

---

[1] H_2S is immediately lethal to man at 1 to 3 mg/litre (650 to 2000 ppm, v/v) of air; at such concentrations it rapidly anaesthetizes the appropriate nasal sensory centres. At 0.24 to 0.36 mg/litre (260 to 240 ppm, v/v) it can be tolerated for 0.5 to 1 hour without ill effect (Karrer, 1960). If a subject becomes intoxicated, he should be removed to fresh air, oxygen should be supplied if necessary, but artificial respiration should only be used if breathing stops. He should be kept warm and rested; qualified medical help should be sought immediately.

for disposal of the effluent gas safely, and the product must be harvested in well-ventilated conditions.

4. All chemostat systems require a gas stream to flow over, or sometimes through, the culture; for *Desulfovibrio*, N_2 is usually used (unless mixotrophy is being studied, when $H_2 + CO_2$ are added) but the special metabolism of the bacteria presents a problem in pH control: evaporation of H_2S leads to alkalinity and altered growth parameters. Excessive chemical buffering can cause osmotic stress and, where feasible, CO_2 should be included in the gas phase, once the culture is in a steady state, to aid pH control.

As in all work with chemostats, media containing precipitates are unacceptable (the reason is that the composition of the medium entering the growth vessel is not defined because the precipitate remains behind). The citrate in medium C will keep the medium components in solution, though a transient turbidity may appear when it is hot. Nitrilotriacetic acid is a relatively non-toxic chelator which may be used when citrate is unacceptable.

Alkalinity accounts for a curious feature of old stagnant cultures of *Desulfovibrio*. Those that have gone alkaline owing to evaporation of H_2S often display a crystalline deposit consisting of magnesium ammonium phosphate; since most media contain Mg^{2+}, NH_4^+ and PO_4^{3-} ions, alkalinity leads to precipitation of the double salt. Because of its relatively uninteresting character, the composition of this precipitate has not been published before. An amusing consequence is that, to my knowledge, at least three workers in this field have independently observed these crystals and analysed them hopefully as an exciting new product before realizing that they had been side-tracked by a piece of simple inorganic chemistry.

Though the primary diagnostic character of sulphate-reducing bacteria is that they grow with sulphate, reducing it to sulphide, they can also grow with sulphite, thiosulphate and, usually, tetrathionate. Postgate (1951a, b) took the view that they do not use elemental sulphur, but Biebl & Pfennig (1977) have questioned this belief (see p. 91). A special physiological group of sulphur-reducing bacteria exists (Pfennig & Biebl, 1981) which does not come within the scope of this monograph. Media containing sulphite were originally used with '*Clostridium*' (*Desulfotomaculum*) *nigrificans* and have been recommended by Mara & Williams (1970) for ordinary sulphate-reducing bacteria (see medium F, Table 2). Most species of *Desulfotomaculum* and some of *Desulfovibrio* can grow with no reducible sulphur compound if an appropriate carbon source is available; four such carbon sources are known at present: pyruvate (Postgate,

1952*a*, 1963*a*), choline (Hayward & Stadtman, 1960), malate or fumarate (Miller & Wakerley, 1966). Such facultative 'non-sulphate' growth yields organisms uncontaminated with sulphide and is in some senses analogous to the fermentative growth of a facultative anaerobe.

Freshly isolated cultures of *Desulfovibrio*, and probably other sulphate-reducing bacteria, sometimes show 'tactophily': a tendency to restrict growth to the walls of the culture tube or the FeS precipitate. In most strains, this property disappears after a few subcultures, although it seems to be a normal feature of the growth of *Desulfonema* species. When it is transient, tactophily seems to be a reflection of a physiological stress inevitably resulting from the transfer of the strain from its natural environment to laboratory culture. Ochynski & Postgate (1963) described the 'training' of a fresh-water strain of *Desulfovibrio vulgaris* to grow in saline media: the first few cultures showed pronounced tactophily which disappeared when the strain was fully adapted.

Salt tolerance

Sulphate-reducing bacteria are found in natural waters of all salinities from near zero to saturation. Above about 2% NaCl the spore-forming types are rarely found and, if they are present, are often not indigenous. Certain genera, species and sub-species have requirements for NaCl which are diagnostic (see Chapter 2). Extreme halophilism in *Desulfovibrio* exists but has not been studied extensively; indeed, the types of sulphate-reducing bacteria which colonize highly saline soils and water need detailed study. Littlewood & Postgate (1957*a*) and Ochynski & Postgate (1963) studied the more moderate halophily, halotolerance and salt sensitivity of marine, brackish water and fresh-water strains of *Desulfovibrio*. They described four types of salt relationship, distinguished by the viable count of populations grown at their habitual salinity and enumerated in media of various salt concentrations. Their results are illustrated diagrammatically in Fig. 7. Fresh-water strains (actually *Desulfovibrio vulgaris*, but not so classified in the early 1960s) were sensitive to NaCl, with only small proportions of the population able to grow above about 1% NaCl and none above about 3% NaCl. Ion replacement studies showed that Na^+ rather than Cl^- was the inhibitory agent. Organisms of saline habit showed the 'intermediate' or 'tolerant' population distributions illustrated in Fig. 7 (the strains tested are now known to have been true *Desulfovibrio desulfuricans*). A type called 'exigent' (now known to be *Desulfovibrio salexigens*, named for this property) had an absolute requirement for more than 0.5 to 1% NaCl and, by ion replacement tests, the required ionic species appeared to be Cl^-, not Na^+.

It is worth drawing attention to the range of salt tolerance of the halotolerant type, strain Canet 41 (Appendix 1), which must be unusual among microbes. A sub-strain was derived from it by 'training' in which the whole population grew readily with 0 to 11% NaCl in the medium.

Tolerance of extreme physical conditions

Thermophily among sulphate-reducing bacteria has been discussed already. Some strains of *Desulfovibrio* are barotolerant and some are also barophilic (ZoBell & Morita, 1957); true barophily seems to be unusual among bacteria (Jannasch, Wirsen & Taylor, 1976) and the sulphate-reducing bacteria may be unique in this respect. Such types have been found in sediments in deep ocean trenches; for details of culturing them

Fig. 7. Classes of salt resistance in *Desulfovibrio*. The diagram illustrates the distribution of salt resistance or requirement in populations of different strains and species of *Desulfovibrio*. Cultures were grown to about 10^9 cells/ml in media resembling medium B (Table 2) with 2.5% NaCl (w/v) added for salt-water strains, and the populations counted in media of various salinities. The freshwater strains (——) were both *Desulfovibrio vulgaris*; the tolerant (– – –) and intermediate (. . . .) strains were *Desulfovibrio desulfuricans*; the exigent strain (— . . — . .) was *Desulfovibrio salexigens*. For further information see text (after Littlewood & Postgate, 1957a).

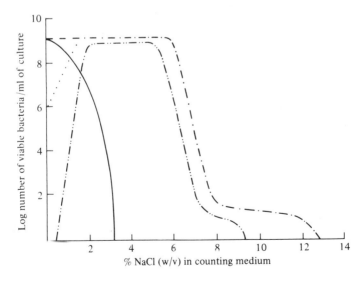

the original literature should be consulted. Psychrophilic sulphate-reducing bacteria exist in nature (Barghoorn & Nichols, 1961) but have not apparently been studied.

Nutrition

Desulfovibrios and desulfotomacula are usually cultured with yeast extract but no special organic growth factors. A need for *p*-aminobenzoic acid and biotin was reported for *Desulfotomaculum ruminis* (Coleman, 1960) and for two out of three strains of *Desulfobulbus propionicus* (Widdel & Pfennig, 1982). *D. acetoxidans* require biotin (Widdel, 1980). Although vitamins are present in medium G (p. 33, based on the formula of Pfennig, Widdel & Trüper, 1981), they are not required absolutely except by a few species. Of three strains of *Desulfobacter postgatei* described by Widdel & Pfennig (1981a), two required *p*-aminobenzoic acid and biotin and one required biotin alone. Reports of vitamin requirements by sulphate-reducing bacteria continue to appear, however, with biotin being the most frequently needed (Skyring, Jones & Goodchild, 1977; Ueki & Suto, 1981); such exigencies may depend on strains as much as species.

A problem in this kind of study is the fact that organic materials such as yeast extract and mixtures of amino acids stimulate growth of *D. vulgaris*, but these effects can be attributed either to an effect on the E_h (by cysteine) or to the chelating action of amino acids on Fe^{2+} (see Postgate, 1965a). This species has no true requirement for such growth factors; they simply assist the uptake of Fe^{2+}, for the organisms have an absolute requirement for relatively high concentrations of iron, though less is needed if the strain grows without reducing sulphate (Postgate, 1956b). The situation regarding iron is interesting because, when sulphate is reduced, all Fe^{2+} is converted to FeS which has an exceptionally low solubility. Neuberg & Mandl (1948) showed that several amino acids and peptides chelate Fe^{2+} and thus inhibit precipitation of FeS, and it is likely that apparent nutritional effects of amino acids and such materials simply reflect their ability to make Fe^{2+} more readily available.

Ammonium ions are the conventional nitrogen source in laboratory media but several strains of *Desulfovibrio* and *Desulfotomaculum* can fix gaseous nitrogen (see p. 96); when they fix nitrogen the molar growth yield (dry weight of organisms produced/mol of carbon substrate consumed in a culture) is much diminished (Senez, 1962; Hill, Drozd & Postgate, 1972).

The range of substrates used by sulphate-reducing bacteria as carbon sources for growth is remarkably narrow: a few C_3- and C_4-substituted fatty acids, such as lactate, pyruvate, fumarate or malate, are widely used, glycerol and certain simple alcohols being less good substrates; some desulfovibrios and desulfotomacula can apparently use carbohydrates such as glucose or sucrose but early impressions that the carbohydrates are widely utilized were false and arose in part from the use of impure cultures. Acetate, propionate and butyrate are not utilized by the lactate-utilizing species of *Desulfovibrio* and *Desulfotomaculum* but species are known which can utilize these fatty acids as well as lactate (see Table 1). *D. sapovorans* is especially interesting because it can utilize long-chain fatty acids; they are oxidized to acetate or propionate (see Chapter 5). Some dispute has centred on whether hydrocarbons, such as methane or higher petroleum hydrocarbons, are utilized for growth; the balance of evidence is that they are not utilized by any known type of sulphate-reducing bacteria. However, ecological evidence suggests that species or consortia capable of such reactions may exist (p. 122).

Mixotrophy

An important feature of the growth of *Desulfovibrio* is its use of 'incomplete' substrates. For example, isobutanol, like gaseous H_2, can support sulphate reduction but not growth. It becomes oxidized to isobutyric acid:

$$2\ \begin{matrix} CH_3 \\ \\ CH_3 \end{matrix}\!\!>\!CH.CH_2OH + SO_4^{2-} \longrightarrow 2\ \begin{matrix} CH_3 \\ \\ CH_3 \end{matrix}\!\!>\!CH.COOH + S^{2-} + 2\,H_2O$$

and tests with labelled isobutanol showed that none is incorporated into the bacteria. Nevertheless, the above reaction yields energy, and this the organism can use to assimilate organic matter, such as the components of yeast extract, and thus grow (Mechalas & Rittenberg, 1960). *Desulfovibrio* growing with isobutanol plus yeast extract resembles in principle *Streptococcus faecalis* growing with glucose plus a variety of growth factors; in that case, too, almost none of the glucose is incorporated. Postgate (1963*b*) described a similar situation with a strain of *Desulfovibrio desulfuricans* able to grow with oxamate; in this case the substrate was degraded to formate, which acted as the 'incomplete' substrate:

$$\begin{array}{ccccccc} \begin{matrix} CONH_2 \\ | \\ COOH \end{matrix} & \xrightarrow{-NH_3} & \begin{matrix} COOH \\ | \\ COOH \end{matrix} & \xrightarrow{-CO_2} & \begin{matrix} H \\ | \\ COOH \end{matrix} & \longrightarrow & \begin{matrix} H_2 + SO_4^{2-} \longrightarrow S^{2-} + H_2O \\ + \\ CO_2 \end{matrix} \\ \text{Oxamate} & & \text{Oxalate} & & \text{Formate} & & \text{Sulphate reduction} \end{array}$$

Fig. 8 illustrates a characteristic of an 'incomplete' substrate: yeast extract is always required for growth in its presence, in contrast to a complete substrate such as lactate, with which yeast extract is only a moderate stimulant.

Coupled assimilation reactions of this kind occur with other organic substrates and could account in part for the vagueness and confusion surrounding the range of substrate specificity among sulphate-reducing bacteria; contaminated cultures may not be the whole story. As a class of microbial nutrition, these coupled assimilation reactions are of great theoretical importance because the substrates are formally acting as sources of inorganic H_2, and H_2 can substitute for them in a purely inorganic (chemotrophic) assimilation process. Rittenberg (1969) named this type of microbial nutrition 'mixotrophy' (as described in Chapter 1, this process earlier led to the delusion that *Desulfovibrio* species were autotrophs).

Fig. 8. Effect of yeast extract on growth of a *Desulfovibrio* species. Cultures with lactate grow without yeast extract but a little augments growth; cultures with oxamate do not grow at all without yeast extract but oxamate permits more growth than does yeast extract alone. *D. desulfuricans oxamicus* grown in media based on □ lactate, ○ oxamate, ◇ neither.

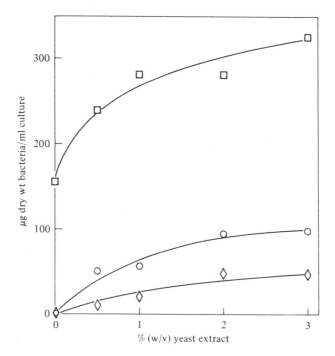

Yeast extract is not the only assimilable material; impurities in laboratory reagents or media can act as assimilable matter and Sorokin (1966a, b, c, d) showed that a mixture of CO_2 + acetate in equimolar proportions could be assimilated in H_2 by one strain. He proposed an energy-requiring carboxylation of acetyl phosphate to pyruvate as the assimilation step, a plausible reaction which requires ferredoxin (see Chapter 5) and has since been demonstrated in the mixotrophy of the Marburg strain of *Desulfovibrio vulgaris* (Badziong, Ditter & Thauer, 1979). Badziong, Thauer & Zeikus (1978) described a medium for isolating mixotrophic strains of *Desulfovibrio*; it seems likely that most strains of *Desulfovibrio* are capable of such mixotrophic growth (Brandis & Thauer, 1981). This type of mixotrophy, which will be discussed further in Chapter 5 (p. 61), has analogies to certain photo-assimilation reactions in photosynthetic bacteria.

The chemotrophic assimilation reactions of *Desulfovibrio* using H_2 were the first examples to be discovered of energy-yielding chemotrophic reactions coupled to an assimilation process; they are now known to occur in hydrogen bacteria, thiobacilli and possibly other bacteria (Rittenberg, 1969). This mixotrophy represents a sort of partial autotrophy which may be of considerable evolutionary significance (see Chapter 6).

Inhibition of growth

An important practical problem is the control of the growth of sulphate-reducing bacteria in economically important situations, such as those discussed in Chapter 8. In consequence, considerable research effort has been devoted to testing various potential microbicides and the results have become scattered throughout the scientific literature. Saleh *et al.* (1964) have compiled an index of growth inhibitors for desulfovibrios and desulfotomacula, a digest of which is presented in Appendix 2. Data for other genera are not available. Inspection of Appendix 2 shows that *Desulfotomaculum* is normally more sensitive to inhibitors in general than *Desulfovibrio*. As with most bacteria, the minimum inhibitory concentrations of microbicides and bacteriostatic substances are usually influenced by the nature of the medium in which the substance is tested, and also by the size of the inoculum, so the quantitative data in the appendix must be taken as 'ball park' numbers. Iron salts can increase the apparent resistance of cultures to inhibitors and, in the case of *Desulfovibrio* species even the presence or absence of NaCl may influence inhibition by, for example, a quaternary microbicide (Costello, King & Miller, 1970).

Despite the quantitative uncertainties, the data in Appendix 2 illustrate the extraordinary resistance *Desulfovibrio*, as a genus, shows to conventional inhibitors such as phenolics, quaternaries, antibiotics and metals. The ineffectiveness of metals is largely due to the ability of the organism to precipitate Hg^{2+}, Cu^{2+}, Cd^{2+}, etc. as sulphides; in the absence of H_2S, these metals are in fact fairly toxic.

Two metabolic inhibitors are available which have a pronounced and therefore specific effect on sulphate-reducing bacteria and are therefore growth inhibitors. The selenate ion is one; it is a competitive antagonist of sulphate reduction which has been valuable in physiological research (see p. 83). The other is the molybdate ion which acts by depleting the organism's ATP pool (Taylor & Oremland, 1979) and which has been used in ecological studies (see Chapter 7).

The resistance of these bacteria to conventional microbicides and microbistats can make their control in economic situations an expensive and uncertain process. Treatments appropriate to specific situations will not be discussed here though some will be mentioned in Chapter 8. Three general principles need restatement, however.

1. *Air is the cheapest and most effective inhibitor.* If any system can be maintained in an aerated condition – even though the dissolved oxygen concentration is vanishingly small – the sulphate-reducing bacteria will not cause trouble. They will not be killed but they will remain dormant. Consequently, in any case of actual or potential economic damage or pollution, the possibilities of aeration should be considered first.

2. *Prevention is infinitely better than cure.* Once established, sulphate-reducing bacteria alter environmental conditions to favour themselves by generating H_2S. This neutralizes useful inhibitors such as chromate (see Chapter 8) and reacts with the oxygen of air, antagonizing its inhibitory effect on the bacteria. The cost of correcting a polluted situation may thus be an order of magnitude or more greater than the cost of foreseeing and preventing it.

3. *Mixed populations can behave aberrantly.* The presence of other bacteria can change the apparent resistance pattern of sulphate-reducing bacteria and this can be inconvenient in practice (Bennett & Bauerle, 1960; Plessis & Gatellier, 1965).

Killing by physical agents

Heat. Temperatures above 50 °C kill mesophilic *Desulfovibrio* spp. within minutes; as with other bacteria, precise survival times depend on the population, environment and rate of heat transfer. Unless the strain has been adapted to high temperatures, it usually dies rapidly at temperatures above 45 °C. Vegetative *Desulfotomaculum* organisms may be comparably heat-sensitive, but sporulation confers heat resistance in bulk populations and studies on the heat sensitivities of the components have not been reported. *Desulfotomaculum nigrificans* spores may withstand the protracted heat sterilization used in food canning (Werkman, 1929), they will survive for 30 minutes at 98–100 °C (Starkey, 1938); *Desulfotomaculum ruminis* spores withstand boiling for 10 minutes; *Desulfotomaculum orientis* spores withstand 10 minutes at 90 °C. Both *D. ruminis* and *D. orientis* may have much greater heat resistances than this, but tests of higher temperatures and longer heating periods have not been reported. Even vegetative *Desulfovibrio* is protected in the dry state: lyophilized ampoules of *Desulfovibrio vulgaris* strain Hildenborough survived 10 minutes at 70 °C in my hands.

Cold. Survival of extreme cold, provided freezing does not occur, is indicated by Barghoorn & Nichols's (1961) report that FeS formation due to sulphate reduction was taking place in an Antarctic pool of some 12% salinity fluctuating between +4 and −40 °C. The taxonomic character of the microbes responsible was not investigated. Freezing suspensions of *Desulfovibrio* in physiological saline or dilute phosphate buffer is an excellent way of obtaining cell-free enzyme preparations (Postgate, 1959*a*, 1963*b*) and kills a considerable proportion of the population. The freezing process is best performed by dropping thick cell suspensions into liquid nitrogen. Glycerol at 10% (v/v) protects populations against freezing damage and may often be used to cold-store populations in a relatively undamaged condition. However, in my experience, some strains show enhanced osmotic fragility after freezing in this way.

Mechanical and sonic damage. Conventional methods of disrupting bacteria are effective with desulfovibrios and desulfotomacula: grinding, decompression and treatment with ultrasonic sound have been recorded for the extraction of enzymes and preparation of membrane fractions.

Desiccation. Freeze-drying may be used to preserve stock cultures providing the drying menstruum is protective (see p. 38). Air, vacuum or acetone drying without protection disrupts the organisms and has been used for obtaining enzymatically active cell preparations.

Visible light. Desulfosarcina variabilis is light sensitive: growth is inhibited completely in ordinary diffuse daylight and the organism must be cultivated in the dark (Widdel, 1980). I am not aware of sensitivity to visible light among other sulphate-reducing bacteria.

Ultraviolet light. Like all living organisms, sulphate-reducing bacteria are sensitive to short ultraviolet radiation. Radiation at 253.7 nm of intensity 44.5 μW/cm^2 at 90 cm caused the count of a saline suspension of *Desulfovibrio vulgaris* strain Hildenborough to fall from 10^8 to 10^2/ml in one minute; 10 minutes' irradiation sterilized the population. Reducing agents had no significant protective effects within the reliability of the counting procedure used, nor did batch cultures and continuous cultures grown at various rates differ significantly (*Chemistry Research*, 1956).

4

Structure and chemical composition

Sulphate-reducing bacteria probably have the same elementary composition as most other bacteria, but their ability to precipitate iron, present in almost all laboratory reagents, probably leads to an excessive iron content in ordinary analyses. Typical analytical data for the Hildenborough strain of *Desulfovibrio vulgaris* (NCIMB 8303, see p. 159) from continuous culture in a medium in which no FeS precipitate was obvious were: C, 46.5%; H, 7.14%; N, 12.47%; S, 1.31%; P, 0.23%.

Littlewood & Postgate (1957*b*) measured the dimensions of *Desulfovibrio vulgaris* strain Wandle both optically and by determining the ion-penetrable space in a known volume of centrifuged cells. The optical measurements gave a mean length of 2.34 μm and a mean diameter of 0.74 μm; considered as cylinders the mean bacterial volume was about 1 μm. Determination of the ion-penetrable space indicated a mean bacterial volume of 1.08 μm^3. Their data enabled them to calculate the bacterial dry weight and other dimensions: one air-dry cell had a mean weight of 0.3125 pg; the mean area of a cell was about 5.5 μm^2; the protein content of the wet bacteria lay between 31% and 35% (w/v). In terms that are more readily visualized, a culture containing 1 mg dry wt of cells would comprise about 3×10^9 individuals having a total bacterial area of 165 cm^2 (approximately one page of this book); 1 ml of packed, centrifuged cells have a total surface area of 3.6 m^2.

Anatomy. Electron microscopy of desulfovibrios or desulfotomacula indicates few unusual details of ultrastructure. The electron micrographs in Chapter 2 (Fig. 4) show flagella originating in a body beneath the cell envelope and the envelopes themselves appear conventionally multilayered. A detailed study was reported by Thomas (1972) on *Desulfovibrio gigas* and *Desulfovibrio vulgaris* (misnamed '*D. hildenborough*'), who also studied lysozyme- and protein-digested preparations. The wall of *D. gigas* is more complex than that of *D. vulgaris* and sections showed fibrillar structures in the cytoplasm as well as large masses of nuclear material. Both species possessed internal membranes which appeared to arise as invaginations of the periplasmic membranes; electron-dense materials (polyphosphates) were present in spheroplasts. Thomas published many elegant micrographs, for which the interested reader should consult his article, but offered little interpretation of the function of the various structures. It seems reasonable to suppose that, like many Gram-negative bacteria, *Desulfovibrio* has an outer membrane, a periplasmic zone and an inner membrane bounding the cytoplasm. A curious observation made by Handley, Adams & Akagi (1973) is that treatment of *D. vulgaris* with mitomycin C induced formation of particles bearing a strong resemblance to bacteriophages. No evidence for infectiousness could be obtained.

Thauer (1982) showed by electron microscopy that *Desulfobacter postgatei* possesses an outer membrane, a cytoplasmic membrane and some seemingly stacked membranes within the cytoplasm.

Evidence for a biochemical anatomy, in the sense that there exists a functional localization of enzymes in *Desulfovibrio* species, arose from the discovery that hydrogenase in *D. gigas* is periplasmic (Chapter 5, p. 71). Such work has been extended by Odom & Peck (1981*a*) for *D. gigas* and by Badziong & Thauer (1980) for mixotrophically-grown *D. vulgaris*. In both organisms, much of the hydrogenase and the *c*-type cytochrome is located outside the protoplast membrane, within the periplasmic space; in *D. gigas*, some formic dehydrogenase also accompanied these substances. They are released by gentle shaking in mildly alkaline buffer or by lysozyme treatment to form 'spheroplasts'. Certain other enzymes and cell components (menaquinone; *b*-type cytochrome; fumarate reductase in *D. gigas*) were localized in the membrane fraction, and yet others (e.g. desulfoviridin, adenylylsulphate reductase, ferredoxin) were largely in the soluble cytoplasmic fraction. Of course, the distribution of these components was not absolute, but the fact that the distribution of the components studied was similar in two different organisms grown in very different cultural conditions signifies that a substantial degree of compartmentation exists in these bacteria. The biochemical anatomy of *Desulfovibrio* was discussed by Peck & Le Gall (1982); its significance will be discussed further in Chapter 5.

Amino acid composition. The amino acid composition of a halophilic strain, probably *Desulfovibrio desulfuricans*, was examined by Subba Rao (1951), using paper chromatography of acid and alkaline hydrolysates of cells. He reported the presence of alanine, arginine, aspartic acid, cystine,

Structure and chemical composition 53

glutamic acid, glycine, histidine, leucine + isoleucine, lysine, methionine, phenylalanine, proline (in traces), serine, threonine, tyrosine, valine. The presence of tryptophan was not definitely established; several 'unknown' spots were observed. Free aspartic acid was detected in the culture fluid. Subba Rao's description of his isolation procedure did not give confidence in the purity of his strain; nevertheless, his observations have been substantially confirmed (Dr Elizabeth Work, personal communication) during an examination of the Hildenborough strain of *Desulfovibrio vulgaris* for diaminopimelic acid. Diaminopimelic acid was present (Work & Dewey, 1953).

Senez (1952) did not detect free amino acids in *Desulfovibrio desulfuricans* strain Canet 41. Ochynski & Postgate (1963) observed a pool of unidentified peptide-like materials in this and other strains which could be extracted into 10% acetic acid. The pool had a primary amino group content equivalent to about 70 μg 'casamino acids'/mg dry wt bacteria. The salt-water habit influenced the character of this pool; desulfotomacula had much lower contents of pool peptides.

Cell walls and spheroplasts. Ochynski & Postgate (1963) reported the preparation of cell walls from *Desulfovibrio vulgaris* and *Desulfovibrio desulfuricans*: they contained numerous amino acids, including diaminopimelic acid, and glucosamine. The organisms are not very sensitive to lysozyme except in the presence of ethylenediaminetetraacetate (EDTA). They recommended 50 μg lysozyme + 400 μg EDTA/ml dilute (1 mM) potassium phosphate buffer at 37 °C for most rapid (about 20 minutes) lysis of cell suspensions. Findley & Akagi (1968) also lysed *D. vulgaris* with lysozyme + EDTA. Lysozyme attacks the 1→4 glycoside linkages of the polysaccharide component of murein, yielding muropeptides. The implication of the lysozyme sensitivity of *Desulfovibrio* is that its walls presumably have a conventional peptidoglycan structure.

Another implication of the presence of diaminopimelic acid is that spheroplasts should be formed by treatment of growing populations with penicillin in an osmotically strong environment. Ochynski & Postgate (1963) described appropriate conditions: spherical, osmotically fragile bodies with residues of walls attached were obtained but the yields were small. The spheroplasts were morphologically quite distinct from 'osmotically fragile vibrios' which they obtained by treatment with lysozyme + EDTA in 0.5–11-M sucrose, potassium phosphate or NaCl as osmotic stabilizer. Ochynski & Postgate (1963) were unable to obtain spheroplasts by this procedure; the osmotically fragile forms retained their vibrio form in osmotically strong media but lysed in dilute saline environments. They were more osmotically stable when prepared from salt-tolerant strains or fresh-water strains which had become acclimatized to a saline environment. Some evidence was presented that the osmotic stability of these bodies, and hence presumably of the parent strains, had to do with the amount and quality of their 'pool' of small peptides. Odom & Peck (1981b) successfully obtained spheroplasts of *Desulfovibrio gigas* by lysozyme + EDTA treatment.

Mucopolysaccharide. Senez (1953) noted that *Desulfovibrio desulfuricans* strain Canet 41, cultured in a yeast extract medium under hydrogen + CO_2, formed sessile growth along the surface of the vessel, associated with an amorphous and viscous substance which he thought to be polysaccharide in nature. Grossman & Postgate (1955) noted slime formation in old cultures of *D. desulfuricans* strain El Agheila Z, and, in continuous cultures of the Hildenborough strain of *Desulfovibrio vulgaris*, cultures have been obtained that were as viscous as white of egg (*Chemistry Research*, 1953). The material was precipitated by ethanol, gave reactions characteristic both of a protein and of a polysaccharide, did not react with iodine and, when hydrolysed with sulphuric acid, yielded only mannose on paper chromatograms (Stacey & Barker, 1960; Ochynski & Postgate, 1963). Its amino acid composition was conventional but quite distinct from cell wall material. The conclusion that it is a mucopolymannoside therefore seems reasonable; Ochynski & Postgate (1963) claimed that its formation was more pronounced among marine strains of *Desulfovibrio* or fresh-water strains acclimatized to saline environments; Stüven (1960) noted that most mucin was formed when his halophilic strain was grown with pyruvate.

Hydrocarbons. Jankowski & ZoBell (1944) first obtained small amounts of unsaponifiable ether-extractable material from heterotrophically grown cultures of *Desulfovibrio desulfuricans*; its analysis corresponded closely to hydrocarbon (see also Chapter 5, p. 69). This may be of importance in considering the genesis of oil in nature. Han & Calvin (1969) regard the hydrocarbons as of primitive character. A review article by Stone & ZoBell (1952) suggested that formation of higher hydrocarbons in traces is a property of several marine micro-organisms besides *Desulfovibrio* (see also Chapter 8, p. 145).

Antibiotics. In 1951 the Hildenborough strain of *Desulfovibrio vulgaris* was screened for antibiotic activity. Filtrates from cultures grown in glucose–peptone and lactate–yeast extract media were tested by the 'cavity plate' assay; no antibiotic activity was detected against *Staphylococcus aureus, Bacillus subtilis, Escherichia coli, Klebsiella pneumoniae, Bacillus cereus, Bacillus mycoides, Pseudomonas fluorescens, Aspergillus niger, Torulopsis utilis* and *Penicillium notatum*.

Tocopherol. Tocopherols are normally plant products but Cinquina (1968) has identified α-tocopherol in a strain of *Desulfotomaculum nigrificans*.

Other components. Proteins, nucleic acids, lipids, etc. of known function or of taxonomic or physiological interest will be described in appropriate sections of this monograph. It is worth reporting that among the antigens in *Desulfovibrio* which give the multiple bands used in taxonomy (Chapter 2, Fig. 5) are cytochrome c_3 and desulfoviridin (Chapter 5). Mysterious molybdoproteins containing iron and sulphur have been isolated and characterized from *Desulfovibrio gigas* (Moura *et al.*, 1976) and *D. africanus* (Hatchikian & Bruschi, 1979); the latter is a decameric protein, mw 112 000, containing Mo_{5-6}, Fe_{20}, S_{20}. Enigmatic cobalt-containing proteins have been isolated from *D. gigas* (Moura, J. J. G. *et al.*, 1980) and from the Norway 4 strain of *D. desulfuricans* (Hatchikian, 1981). Both are low molecular weight porphyroproteins, the latter being slowly reducible by hydrogenase + cytochrome c_3. Catalases and superoxide dismutases are present in *Desulfovibrio vulgaris, D. gigas* and the Norway 4 strain of *D. desulfuricans*, but not in the type strains of *D. desulfuricans* and *D. salexigens* (Hatchikian, Le Gall & Bell, 1977); the superoxide dismutase extracted from strain Norway 4 (Bruschi *et al.*, 1977) is unusual in having iron as its metal centre, not the copper or zinc usually found in bacteria, nor the manganese of most prokaryotes. The physiological significance of the presence of such 'enzymes' in so strict an anaerobe is perplexing and it is tempting to take the view, expressed by some for superoxide dismutase in general, that their O_2-related activities are trivial functions of metalloproteins whose real biological functions are not yet known (Fee, 1982).

5

Metabolism

Some theoretical data

The results of thermodynamic calculations of the energy relations in reactions known or presumed to be conducted by sulphate-reducing bacteria have no detailed biological significance but can be a useful guide to their metabolic patterns. Baas-Becking & Parks (1927), Rittenberg (1941), Ishimoto, Koyama & Nagai (1955), Sorokin (1957), Wake et al. (1977), Thauer, Jungermann & Decker (1977) and Wagner, Kassner & Kamen (1974) quoted certain data; as the latter authors pointed out, the numerical results depend on the standard data the author chooses to use. Enthalpy (ΔH) data were calculated for cultures of four strains of *Desulfovibrio* grown with lactate or pyruvate and showed reasonable agreement with experimental values obtained by microcalorimetry (Traore, Hatchikian, Belaich & Le Gall, 1981; Traore, Hatchikian, Le Gall & Belaich, 1982).

Table 3 gives the results of calculations from standard data of theoretical values for a number of reactions at 27 °C. Values for ΔH_f^0 of the pure acids were calculated from heat of combustion data (Hodgman, 1949) and assumed equal to ΔH_f^0 of the aqueous ions; ΔG_f^0 values for inorganic ions were obtained from tables (Rossini et al., 1952) except that a corrected value for the thiosulphate ion (Mel, 1954) was used; ΔG_f^0 values for the ions of organic acids were taken mainly from Burton & Krebs (1953); Burton's (1955) correction did not apply to any of the substances considered. The columns in Table 3 are: ΔH^0, the standard heat of reaction for the equation as written; $\Delta H^0/H_2$, the standard heat content change for each two-electron transfer; ΔG^0, the standard free energy of the reaction as written; $\Delta G^0/H_2$, the standard free energy change for each two-electron transfer. These figures are artificial from a biological point of view, since at pH 7 the sulphide ion is largely hydrolysed, and that hydrolysis yields considerable free energy. The columns headed $\Delta G'$ and $\Delta G'/H_2$ list the

Table 3. Thermodynamic data for certain biochemical reactions of sulphate-reducing bacteria (in kJ)[a]

Reaction	ΔH^0	$\Delta H^0/H_2$	ΔG^0	$\Delta G^0/H_2$	$\Delta G'$	$\Delta G'/H_2$
$4H_2 + SO_4^{2-} = S^{2-} + 4H_2O$	-196.46	-48.91	-123.98	-30.93	-172.05	-42.84
$3H_2 + SO_3^{2-} = S^{2-} + 3H_2O$	-191.44	-63.95	-134.60	-44.73	-182.67	-60.61
$4H_2 + S_2O_3^{2-} = S^{2-} + H_2S + 3H_2O$	-210.67	-51.00	-145.88	-36.53	-193.70	-47.44
$9H_2 + S_4O_6^{2-} = S^{2-} + 3H_2S + 6H_2O$	-621.15	-56.01	-401.70	-44.73	-449.77	-50.16
$S_2O_3^{2-} + H_2 = SO_3^{2-} + H_2S$[b]	-19.23	-19.23	-10.87	-10.87	-10.87	-10.87
$Acetate^- + SO_4^{2-} = H_2O + CO_2 + HCO_3^- + S^{2-}$	$+48.07$	$+12.12$	-12.41	-3.09	-60.48	-15.13
$3\,acetate^- + 4SO_3^{2-} = 3H_2O + 3CO_2 + HCO_3^- + 4S^{2-}$	-12.54	-0.79	-171.38	-14.21	-363.66	-30.10
$CH_4 + SO_4^{2-} = CO_2 + 2H_2O + S^{2-}$[c]	$+40.55$	$+10.11$	$+16.72$	$+4.18$	-31.77	-7.94
$2C_2H_5OH + SO_3^{2-} = 2\,acetic\ acid + 2H_2O + S^{2-}$[d]	-55.59	-13.92	-59.36	-20.06	-128.33	-32.19
$4\,formate^- + SO_4^{2-} = 4HCO_3^- + S^{2-}$	-213.60	-53.50	-182.67	-45.56	-230.74	-57.68
$4\,pyruvate^- + SO_4^{2-} = 4\,acetate^- + 4CO_2 + S^{2-}$	-351.12	-87.78	-331.06	-82.76	-379.13	-94.89
$2\,lactate^- + SO_4^{2-} = 2\,acetate^- + 2CO_2 + 2H_2O + S^{2-}$[e]	-79.42	-19.85	-140.45	-35.11	-188.52	-47.02
$2\,malate^{2-} + SO_4^{2-} = 2\,acetate^- + 2CO_2 + 2HCO_3^- + S^{2-}$	$+154.66$	$+38.66$	-181.00	-45.35	-229.06	-57.06
$2\,fumarate^{2-} + 2H_2O + SO_4^{2-} = 2\,acetate^- + 2CO_2$ $+ 2HCO_3^- + S^{2-}$	$+154.66$	$+38.66$	-190.19	-47.44	-238.26	-59.56
$4\,succinate^{2-} + 3SO_4^{2-} = 4\,acetate^- + 4CO_2 + 4HCO_3^- + 3S^{2-}$	$+305.14$	$+6.35$	-150.48	-12.54	-294.69	-24.66

[a] These values have been converted from calories to Joules because the latter are now the International System (SI) units of energy. For comparison with data published earlier, 1 cal = 4.18 J.
[b] Ishimoto et al. (1955) demonstrated this reaction in a cell-free preparation; they quoted a calculated value of $\Delta G = +0.59$ kJ but gave no reference to the standard data used for the calculation.
[c] Sorokin (1957) quoted several positive values of ΔG^0 for this reaction but did not consider values for $\Delta G'$: his conclusion that the reaction is thermodynamically impossible was therefore incorrect for pH 7.
[d] $\Delta G'$ values not corrected for ionization of acetic acid.
[e] Rittenberg (1941) calculated $\Delta G' = -178.07$ kJ for this reaction.

free energies of reaction corrected to pH 7. This correction was made using the first and second dissociation constants of H_2S (Hodgman, 1949), calculating the free energy of the hydrolysis:

$$S^{2-} + H^+ \rightarrow \underbrace{H_2S + HS^- + S^{2-}}$$
$$[1\text{ M}] \quad [10^{-7}\text{ M}] \quad [1\text{ M}]$$

for which $\Delta G = -48.1$ kJ/g ion S^{2-}, and subtracting this from the values of ΔG^0. $\Delta G'$ and $\Delta G'/H_2$ thus represent the free energy change with all reactants in their standard state except sulphide, which is at equilibrium at pH 7. Table 3 illustrates a few points of interest. Some of the reactions are endothermic (ΔH^0 is positive) but exergonic (ΔG^0 is negative): Wilson & Peterson (1931) noted similar reactions when calculating the energetics of fermentations of several other bacteria. The free energy of the oxidation of acetate to CO_2 is low, which may account for the widespread appearance of acetate as an end-product of the metabolism of dissimilatory sulphate-reducing bacteria. Similarly, the apparent absence of methane-utilizing strains (Sorokin, 1957, see also p. 122) can be attributed to the low energy yield to be expected therefrom. Succinate and ethanol are known to give poor cell yields in culture, as might be expected from their low $\Delta G'/H_2$ values. Aromatic compounds, such as benzoic acid or benzene itself, give extremely low $\Delta G'$ values (not quoted in Table 3), whether the end-product is taken as acetate or CO_2; hence, as Novelli & ZoBell (1944) reported, these compounds would be poor hydrogen donors for sulphate reduction. Nevertheless, benzoate-oxidizing sulphate-reducing bacteria have been isolated (Widdel, 1980).

Wake *et al.* (1977) concluded that a theoretical energy yield of at least 35 kJ for every two electrons transferred was necessary to support growth of *Desulfovibrio* or *Desulfotomaculum*. However, the existence of acetate- and benzoate-utilizing sulphate-reducing bacteria suggests that this conclusion is only valid for relatively fast-growing sulphate reducers.

The standard free energy of reaction is simply related to the standard oxidation–reduction potential (E'_0) of the reaction; certain of the values of E'_0 at pH 7 calculated from the values given in Table 3 are of interest because they indicate the sort of potential at which electron transfer takes place in the cellular respiratory processes. Potentials for the reduction of various substrates in hydrogen are given in Table 4, derived from the relationship $-\Delta G' = nFE_0$. Wagner *et al.* (1974) obtained data in a similar range and published E_0 values for partial reductions such as trithionate formation (-0.355 mV), reduction to sulphite only (-0.442 mV) and the reductive dismutation of thiosulphate to sulphide and sulphite

Table 4. *Standard potentials (in mV) calculated for certain reductions conducted by* Desulfovibrio desulfuricans

Reaction	$\Delta G'$ (kJ)	E_0' (pH 7)
$4H_2 + SO_4^{2-} = S^{2-} + 4H_2O$	−172.0	−188
$3H_2 + SO_3^{2-} = S^{2-} + 3H_2O$	−182.7	−96
$4H_2 + S_2O_3^{2-} = S^{2-} + H_2S + 3H_2O$	−193.7	−162
$9H_2 + S_4O_6^{2-} = S^{2-} + 3H_2S + 6H_2O$	−449.76	−152

(−0.390 mV). They also listed potentials derived from other standard data.

Despite numerical differences, calculations of the kind shown in Table 4 illustrate the highly reducing potential at which many metabolic reactions of sulphate-reducing bacteria take place, particularly when compared with the reduction of nitrate in hydrogen to ammonia:

$$NO_3^- + 4H_2 + 2H^+ = 3H_2O + NH_4^+;$$
$$\Delta G' = -149.2 \text{ kJ/H}_2; E = 363 \text{ mV}$$

The sulphate-reducing bacteria conduct their major metabolic oxidations within a redox span whose positive end is around E_h −150 to −200 mV; this may account for the strictly anaerobic habit of the organisms since they need not possess electron-transporting co-factors which would have an appreciable proportion of their reduced form stable at positive E_h values. Their growth is accompanied by a drop in redox potential to the range of −250 mV, which is not surprising since the E_h values of solutions of Na_2S fall in this range. But there is evidence that a drop in potential to the range of −200 mV is necessary for the *initiation* of growth (see Chapter 2).

Broad metabolic patterns

The metabolism of ordinary aerobic bacteria can be broadly divided into anabolic (biosynthetic) and catabolic (dissimilatory) processes.[1] Almost nothing is known of anabolic processes in sulphate-reducing bacteria and

---

[1] A third class, anaplerotic processes, concerns reactions whereby monomers for biosynthetic processes are generated (Kornberg, 1966); it is not relevant in the present context

this aspect of their metabolism will feature only rarely in this chapter. Catabolic processes in aerobes can be divided into the three stages illustrated in Fig. 9: the breakdown of carbon substrates is essentially an anaerobic process during which a small amount of 'substrate level' ATP may be generated but, more important, the primary electron donors to the electron transport chain (usually NAD or NADP, sometimes flavoproteins) become reduced. Electron transport, also an anaerobic process, leads to relatively efficient generation of ATP ('respiratory chain phosphorylation') and only at the final stage, reduction of oxygen to OH^+ by cytochrome oxidase, does a truly aerobic reaction take place. Most facultative anaerobes can omit the electron transport and oxygen reduction stages and then rely solely on substrate level phosphorylation for their energy budget; in this respect they resemble the fermentative obligate anaerobic bacteria, such as the clostridia.

Dissimilatory sulphate reduction is essentially an oxidative type of metabolism, despite the obligate anaerobic habit of the organisms concerned, and the division into two non-oxidative processes, dissimilation and electron transport, followed by an oxidative process, sulphate reduction, can usefully, though not universally, be applied. Fig. 10 illustrates in broad outline the framework in which the metabolism of *Desulfovibrio* will be discussed; notice that catabolism ceases at the acetate level. It is likely that this framework does not apply to desulfotomacula (p. 94).

In addition to carbon sources, gaseous hydrogen can act as an electron

Fig. 9. A formalized scheme for aerobic respiration. Both carbon catabolism and electron transport are anaerobic processes. The only oxidative step is the oxidation of the terminal oxidase by oxygen, amounting to a reduction of oxygen to OH^- ions. ATP is 'gained' both at the substrate level and during electron transport.

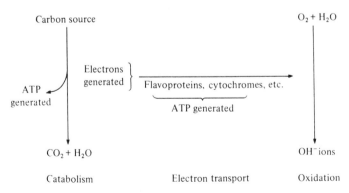

CO_2 assimilation

donor for the electron transport chain of *Desulfovibrio*, a reaction which does not occur in aerobic microbes except among certain specialized bacteria called the hydrogen bacteria ('*Knallgassbakterien*').

This chapter will be concerned with all three stages of the metabolism illustrated in Fig. 10 as well as with the few assimilatory processes studied in these organisms. Most studies of the metabolic chemistry of sulphate-reducing bacteria have been made on species of *Desulfovibrio*, so the emphasis will necessarily be on this genus. Where species differences seem important, these will be mentioned and, as far as possible, comparable information covering *Desulfotomaculum* and other genera will be included. But this is still a developing subject, advancing rapidly, and I have chosen a conservative approach even at the risk of simplifying complex questions. For the most up-to-date information, recent reviews and the primary literature must be consulted.

CO_2 assimilation

The incorrect belief that *Desulfovibrio* strains can grow autotrophically was discussed in Chapter 1 (p. 7): the experimental data on which it was based – apparent enhancement of growth in putatively mineral media when $H_2 + CO_2$ were provided – is now attributed to mixotrophy. The organisms are capable of generating energy by the anaerobic oxidation of hydrogen, a lithotrophic process, but pre-formed organic matter, acetate in particular, is required for growth and CO_2 assimilation. As mentioned

Fig. 10. A formalized scheme for dissimilatory sulphate reduction. Compared with Fig. 9, the broad pattern is similar but carbon catabolism ceases at the acetate level of oxidation and the oxidative step involves removal of oxygen atoms from sulphate and its reduction to sulphide. ATP is lost at this stage.

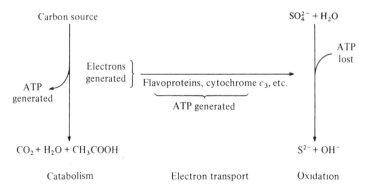

62   Metabolism

in Chapter 3 (p. 47), Sorokin (1966a, b, c, d) demonstrated that a strain of *Desulfovibrio* could assimilate equimolar acetate + CO_2 at the expense of H_2 oxidation with sulphate. Several strains of *Desulfovibrio* seem able to grow by this mechanism (Brandis & Thauer, 1981); in *D. vulgaris* strain Marburg, which grows especially readily in these conditions, Badziong, Thauer & Zeikus (1978) showed that 70% of the cell carbon came from acetate and 30% from CO_2 in these mixotrophic conditions. Sorokin had suggested that reductive carboxylation of acetate (as acetyl phosphate) to pyruvate was the pathway of acetate and CO_2 assimilation, but presented no substantial evidence for this view. Suh & Akagi (1966) demonstrated this reaction and showed that it required ferredoxin, but it rested with Badziong, Ditter & Thauer (1979) to demonstrate that it is actually the assimilation pathway in *D. vulgaris*. Alvarez & Barton (1977) had earlier suggested that, because of the presence of ribulose bis-phosphate (RUBP) carboxylase and phosphoriboisomerase in *D. vulgaris*, a sequence approximating to the Calvin cycle was the pathway of CO_2 assimilation; if the true pathway is indeed carboxylation of acetate, which seems plausible, then the apparent presence of a RUBP carboxylase system is enigmatic.

Biochemical studies of mixotrophic, or perhaps even autotrophic, CO_2 assimilation in the newer genera of sulphate-reducing bacteria had not been reported by late 1982.

Carbon dissimilation

Knowledge of the catabolic pathways in sulphate-reducing bacteria is somewhat fragmentary, partly because they are rather recalcitrant research subjects, and partly because more exotic aspects of their metabolism have attracted greater attention. The historical reasons for this shortage of hard information were mentioned in connection with the classification of sulphate-reducing bacteria (Chapter 2) but they can usefully be summarized again here.

1. The criteria of purity of strains used have not always been satisfactory. Commensal organisms can cause misleading results; several examples exist of cultures able to utilize specific carbon sources only when impure.

2. Growth of a strain with an unfamiliar carbon source may require a metabolic adjustment which delays growth. Growth may be prevented altogether unless the redox potential of the medium is artificially poised by adding thioglycollate or sulphide (cysteine is unsuitable for the purpose since it acts as a source of carbon).

Carbon dissimilation

3. This metabolic adjustment (which may be a permeability change, derepression of enzyme formation or conceivably selection of mutants) is assisted by the presence of yeast extract (Pochon & Chalvignac, 1952). Tests will thus differ in their results according to whether they were made in the presence or absence of yeast extract.

4. The dehydrogenase pattern, as indicated by ability to reduce dyestuffs or utilize substrates in the Warburg manometer, is much influenced by the previous history of the strain. 'Adaptive' response of this kind to glucose (*Chemistry Research*, 1952, p. 93), dicarboxylic acids (Senez, 1954) and malate (Grossman & Postgate, 1955) have been recorded, yet it is difficult to believe that at least some of these substrates do not normally participate in the metabolism of the strains examined.

5. Certain substrates (e.g. citrate) prevent the precipitation of FeS until high sulphide concentrations are reached and can give false negative results if iron sulphide formation is taken as a criterion of growth. In contrast, however, substrates such as ethanol permit very poor growth though the yield of sulphide is high, and the question arises whether they can usefully be described as substrates for growth. They are on the borderline of being mixotrophic substrates like isobutanol or gaseous hydrogen (see Chapter 3, p. 45).

It is therefore not practicable to compose a metabolic scheme for the breakdown of even such simple carbon compounds as have been studied. A descriptive account of their metabolism follows.

Lactate is oxidized by *Desulfovibrio*, probably *via* pyruvate, to yield acetate and CO_2 as major end products. Lactate dehydrogenase activity is often difficult to demonstrate in extracts and activity can be sensitive to O_2 (Stams & Hansen, 1982), but a stable D-lactate dehydrogenase has been purified from *D. vulgaris* by Ogata, Arihara & Yagi (1981) and appears to transfer electrons to a specific cytochrome (see below). Peck & Le Gall (1982) mentioned the existence of a distinguishable L-lactate dehydrogenase in *D. gigas*. Both L- and D-lactate dehydrogenases are located in the cytoplasm; if both L- and D-enzyme are present in the same organism, a racemase would not be necessary. With living cells of *Desulfovibrio*, no gross lactate metabolism occurs in the absence of sulphate, but there is some evidence that H_2 can be formed from lactate in small amounts, a reaction which is 'pulled' by constant removal of hydrogen (Bryant *et al.*, 1977; a matter to be discussed shortly). Certainly hydrogen could well be an intermediate in lactate metabolism; the enzyme hydrogenase is necessary for lactate oxidation because protoplasts of *D. gigas*, which lack soluble hydrogenase and cytochrome c_3, cannot oxidize lactate with sulphate unless they are provided with both of these substances (Odom & Peck, 1981b). The hydrogenase is probably involved in proton transfer from lactate across the cell membrane (see pp. 94–95).

Desulfobulbus propionicus can grow in lactate media without sulphate; lactate is fermented to a mixture of acetate and propionate (Widdel & Pfennig, 1982). This species requires sulphate to grow with propionate, from which it forms acetate.

Lactate, pyruvate, glycerol, ethanol and the acids of the tricarboxylic acid cycle are all converted to acetate and CO_2 as major end-products by *Desulfovibrio* (see, for example, Senez, 1954; Grossman & Postgate, 1955). In this genus, some incorporation of acetate may occur as a result of mixotrophic assimilation (p. 46), but further oxidation of acetate coupled to sulphate reduction does not occur in most species (see Table 1). Homologues of ethanol (propanol, butanol, etc.) are metabolized to the homologous fatty acid (propionic, butyric) and are often 'incomplete' mixotrophic substrates.

Pyruvate can usually be metabolized in the absence of sulphate to yield $H_2 + CO_2$ and acetate, a reaction discussed further in the next section. Lactate and ethanol show no such reaction in pure culture but the ability of desulfovibrios to grow in sulphate-free culture with an H_2-consuming methanogen led Bryant *et al.* (1977) to propose that they could generate H_2 from these substrates if another organism scavenged it forthwith. Tsuji & Yagi (1980) showed that cultures of *D. vulgaris* evolved H_2 at an early stage of growth but whether lactate was the true source of the H_2 was not clearly established.

Cysteine is desulfhydrated and metabolized to acetate (Senez & Leroux-Gilleron, 1954); choline (the hydroxyethyltrimethylammonium ion), when it is metabolized at all, is also metabolized to acetate probably via acetaldehyde, like ethanol, with trimethylamine as a by-product. A cell-free enzyme system has been prepared and studied (Hayward & Stadtman, 1959, 1960; Hayward, 1960).

The C_1 compound formate and the C_2 compounds oxalate and oxamate can be completely metabolized to CO_2, though the growth yields are poor and it is doubtful whether any are 'complete' substrates. Growth with methanol has not been established rigidly with *Desulfovibrio*[1]. Formate has been recommended as a selective substrate for enrichment cultures (Sorokin, 1954a). Formate is also metabolized, but without growth,

---

[1] Methanol was said to be oxidized in the first edition of this book (p. 53) on the basis of an unpublished survey which may not have been reliable.

Carbon dissimilation 65

in the absence of sulphate, to $H_2 + CO_2$. Formic dehydrogenases of differing properties have been isolated by Yagi (1969) and Riederer-Henderson & Peck (1970) from strains of *Desulfovibrio*. Formic dehydrogenase is periplasmic: located outside the cell membrane (see Peck & Le Gall, 1982). The oxidation of oxamate or oxalate, which takes place via formate, was mentioned in Chapter 3 (p. 45). Yagi (1958, 1959, also Yagi & Tamiya, 1962) discovered an interesting reaction in extracts of their bacteria: the oxidation of carbon monoxide, which probably proceeds via formate:

$$CO \xrightarrow{H_2O} H.COOH \xrightarrow{SO_3^{2-}} H_2O + CO_2$$

but whether growth at the expense of CO oxidation can take place is uncertain.

Utilization of carbohydrates by sulphate-reducing bacteria is still a matter of dispute; Postgate (1959*a*) discussed how problems of contaminated cultures (Chapter 1) and effects of yeast extract (now recognized as permitting growth with incomplete substrates: Chapter 3), as well as inattention to the E_h of the culture, could give misleading results in laboratory tests. As a general rule, pure cultures of *Desulfovibrio* do not readily metabolize these substrates, but exceptions exist. Glucose is definitely utilized by some strains of the thermophile *Desulfotomaculum nigrificans* and Akagi & Jackson (1967) have provided evidence that the gross products, acetate, CO_2 and ethanol, arise by the Entner–Doudoroff pathway.

Enzymes of the conventional tricarboxylic acid cycle have been sought in *Desulfovibrio* and many are present, but Lewis & Miller (1975, 1977) concluded that the cycle is incomplete. They believed that the cycle has no catabolic function though sufficient enzymes may have been present, below the limits of detection, for it to perform an anaplerotic function. In some strains the citrate synthase is unusual in that the acetyl groups form the 1 and 2 carbon atoms of citrate instead of the usual 5 and 6 after the condensation of acetyl coenzyme A and oxaloacetate (Gottschalk & Barker, 1967); a dispute exists about the significance of this observation (Dittbrenner, Chowdury & Gottschalk, 1969; O'Brien & Stern, 1969). Domka & Szulczynski (1981) showed that *D. vulgaris* oxidizes glycerophosphates to acetate and CO_2.

Despite uncertainty about the existence and function of the tricarboxylic acid cycle, there is little doubt that a cyclic terminal carbon metabolism does take place in *Desulfovibrio*. Grossman & Postgate (1955) showed that *D. desulfuricans* coupled processes such as malate oxidation to the reduction of fumarate to succinate, and proposed a linked cycle for sulphate reduction:

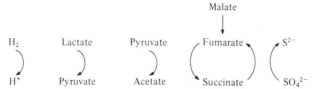

Today such a mechanism would have to be expressed in a much more sophisticated manner if only to incorporate the electron transport factors (cytochromes c_3, ferredoxins, etc. – see p. 73–82) but the evidence on which it was based still stands. Succinate and fumarate accumulate transiently during oxidation of malate, and fumarate can act as a substitute electron acceptor, replacing sulphate for hydrogen or lactate oxidation (even supporting sulphate-free growth of some strains: Miller & Wakerley, 1966; Miller, Neumann, Elford & Wakerley, 1970). Fumarate is reduced to succinate. Catalytic concentrations of fumarate (or other intermediates such as pyruvate or lactate) accelerate sulphate reduction in hydrogen. Fumarase (the enzyme conducting the reversible hydration of fumarate to malate) is present in *Desulfovibrio desulfuricans* (Grossman & Postgate, 1955) and so is a malic enzyme in *D. gigas* which converts malate directly to pyruvate (Hatchikian & Le Gall, 1970a, b). A central role for fumarate in the terminal metabolism of *D. gigas* is impressively illustrated by the demonstration that respiratory chain phosphorylation can accompany the reduction of fumarate by hydrogen or by lactate (Barton, Le Gall & Peck, 1970; Barton & Peck, 1971). Fumarate–succinate cycles are probably involved in energy generation processes in other anaerobes as well as sulphate-reducing bacteria (see Gottschalk & Andreesen, 1979).

Pyruvate metabolism

Pyruvic acid – or rather pyruvate – also plays a central part in the metabolism of these bacteria. Lactate is oxidized via pyruvate to acetate. Desulfovibrios, desulfobacters and desulfotomacula all possess the pyruvic phosphoroclastic system in which, in the absence of sulphate, pyruvate is dismuted to acetyl phosphate, CO_2 and hydrogen. The system is located within the cell cytoplasm (see Peck & Le Gall, 1982). A curious feature of the reaction is that, in *Desulfovibrio*, it is stimulated by ATP (Yates, 1967), a property shared by certain clostridia but not by other bacteria.

Carbon dissimilation 67

Requirements for activity include thiamine pyrophosphate, Mg^{2+} and coenzyme A (Millet, 1954) and, in *Desulfovibrio vulgaris*, either ferredoxin or cytochrome c_3 is needed for activity; the two carriers act synergistically so that optimal activity is only obtained with both together (see Akagi, 1967). Acetokinase is present in the organism. A minor product of pyruvate phosphoroclasm is methane, the formation of which involves vitamin B_{12} and ATP (Postgate, 1969b); Sadana (1954) reported ethanol as a major product of pyruvic dismutation by intact *D. vulgaris* but this observation has not been confirmed in other laboratories. Kadota & Miyoshi (1964) reported traces of formate in cultures grown on pyruvate.

The ability of pyruvate to support sulphate-free growth of certain species of *Desulfovibrio* and *Desulfotomaculum* was mentioned in Chapter 3. The gross products of growth in cultures are acetate, CO_2 and H_2, so the reaction is obviously a typical phosphoroclasm leading to the generation of acetyl phosphate and hence ATP. The reaction is found in most species: even where pyruvate alone will not readily support growth, as in *Desulfovibrio vulgaris*, suspensions of cells dismute pyruvate to acetate, CO_2 and H_2. Where pyruvate dismutation supports growth, it provides the bacteria with a 'sulphate-free' mode of growth analogous to the 'oxygen-free' growth of a facultative aerobe; it may thus have survival value for such species in sulphate-deficient environments. *Desulfobulbus propionicus* can grow on pyruvate (or lactate) without sulphate (Widdel & Pfennig, 1982).

As in several other anaerobes, the pyruvic phosphoroclastic system can function in reverse, CO_2 being 'fixed' by reductive carboxylation to pyruvate. The reaction requires ferredoxin (Suh & Akagi, 1966) and may well participate in the 'mixotrophic' metabolism of *Desulfovibrio*. It is also involved in acetate metabolism by *Desulfobacter postgatei* (below).

Acetate metabolism

The discovery of several species of sulphate-reducing bacteria which are capable of metabolizing acetate immediately raises questions about the pathway of acetate utilization. A summary of the first study of this kind was presented by Thauer (1982): in *Desulfobacter postgatei* it seems that a conventional citric acid cycle operates. Evidence was obtained for high activities of citrate synthase, aconitase, isocitrate dehydrogenase, α-ketoglutarate dehydrogenase, succinic dehydrogenase, fumarase and malate dehydrogenase. Some of these enzymes have unusual properties: the fumarase is cold-labile, the citrate synthase is AMP-dependent, the malate dehydrogenase couples with quinones rather than pyridine nucleotides;

68 *Metabolism*

but it seems likely that a dissmilatory citric acid cycle operates. For the formation of intermediates for biosynthesis (anaplerosis, see p. 59), a pyruvate synthase system, conducting the ferredoxin-dependent reductive carboxylation of acetyl coenzyme A to pyruvate, probably operates. Thauer (1982) mentioned that a comparable metabolic pathway may well operate in the acetate-utilizing sulphur-reducing organism *Desulfuromonas acetoxidans*; in both organisms the relatively oxidizing potential of the succinate→fumarate step implies that some energy input will be required at this stage of the cycle when it is coupled to reactions of relatively narrow redox span such as sulphate or sulphur reduction.

Organisms able to oxidize acetate are often unable to metabolize lactate or pyruvate (e.g. *Desulfotomaculum acetoxidans*, *Desulfobacter postgatei*) although most (but not *D. postgatei*) can oxidize ethanol or butyrate completely to CO_2 (e.g. Widdel & Pfennig, 1977). The reason for this apparent metabolic incompatibility is not entirely clear.

Metabolism of other carbon sources

The present consensus view is that hydrocarbons are not attacked by sulphate-reducing bacteria (see p. 45) although methane seems to disappear from zones of sulphate reduction in sediments (see Chapter 7). No biochemistry has been reported on the metabolism of benzoate, propionate and other substrates by those species capable of attacking them, but Pfennig & Widdel (1982) reported that *Desulfovibrio sapovorans*, which degrades higher fatty acids, forms exclusively acetate from C-even fatty acids but forms equimolar acetate plus propionate from C-odd fatty acids. They took this as evidence that the classical β-oxidation of the aliphatic carbon chain was taking place.

Products of carbon metabolism

Apart from cells themselves, which arise in relatively low specific yield compared with other bacteria, the end-products of carbon oxidation coupled to sulphate reduction by most *Desulfovibrio* species are acetate, water and, usually CO_2. Oxidation of ethanol and higher alcohols yields no CO_2 and the fatty acid need not be acetic. Formate, oxamate and oxalate, when utilized, yield CO_2 exclusively. In the absence or near absence of sulphate, hydrogen, ethanol and other fragments of the substrate may accumulate and with some substrates hydrogen may become a major metabolic product in sulphate-deficient conditions (see above).

Carbon dissimilation

Thus the gross products of carbon metabolism are reasonably predictable in *Desulfovibrio*. In the acetate-utilizing species, CO_2 is the major product, but their metabolism has been less extensively studied.

In *Desulfovibrio*, two minor products of carbon metabolism are formed which should be mentioned.

Hydrocarbons. Methane formation in cultures of sulphate-reducing bacteria is well established. Postgate (1969*b*) surveyed earlier reports. Sisler & ZoBell (1951*a*), growing marine *Desulfovibrio desulfuricans* in putatively mineral media, observed an uptake of hydrogen in excess of the theoretical amount required for sulphate reduction. A few of their cultures produced methane but the quantity was small; other products included lipids and hydrocarbons. A 16-litre culture of one strain produced, after 50 days at 32 °C, 0.72 g of CCl_4-soluble material of which nearly 30% was an unsaponifiable oil consisting most probably of paraffinic or naphthenic hydrocarbons. Carbon dioxide corresponding to an $H_2 : CO_2$ ratio of 2.37 disappeared from the system during growth, consistent with synthesis of a highly reduced product. Oppenheimer (1965) observed hydrocarbon formation by anaerobic bacteria particularly under pressure, including a pure strain of *Desulfovibrio desulfuricans*. Hydrocarbon formation was mentioned on p. 54.

Mucin. A mucopolysaccharide based on mannose appears in old cultures of *D. desulfuricans* and *D. vulgaris* and was discussed in Chapter 4.

Oxidative carbon metabolism without participation of sulphate

Sulphite, thiosulphate or tetrathionate can substitute for sulphate as electron acceptors for growth and carbon metabolism. Fumarate will also support oxidative metabolism (above). By virtue of the autoxidizibility of native cytochromes c_3, certain electron acceptors unrelated to sulphate may permit metabolism though not growth. Obvious examples are redox dyes such as methylene blue or viologens. Pichinoty & Senez (1958) demonstrated pyruvate oxidation via cytochrome c_3, using nitrite as the electron acceptor. By analogy with the Knallgass reaction (p. 61), oxygen can serve as terminal acceptor for carbon metabolism mediated by cytochrome c_3, and I have observed quantitative oxidation of lactate, pyruvate and malate to acetate with air as the electron acceptor. It is curious how closely *Desulfovibrio* approaches a facultative aerobic metabolism without, apparently, being able to channel this metabolism into biosynthesis and growth. Some strains of *Desulfovibrio* can apparently grow by lactate oxidation coupled to nitrate reduction. Nitrate reduction is discussed later in this chapter (see p. 96).

Fractionation of carbon isotopes. Consistent with the fractionation of sulphur isotopes during sulphate reduction (see p. 89), a modest separation of carbon isotopes ^{12}C and ^{13}C also occurs. Kaplan & Rittenberg (1964b) showed that, with lactate as carbon source and sulphite as terminal electron acceptor, the slower the bacterial metabolism, the higher the proportion of ^{12}C obtained in the cells and CO_2.

Hydrogen metabolism

A striking feature of the carbon metabolism of *Desulfovibrio* is the involvement of gaseous H_2 at several stages. Hydrogen metabolism seems to be central to the metabolism of carbon compounds, as well as supporting the mixotrophic metabolism described earlier in this chapter. Clear examples of the involvement of hydrogen are provided by the pyruvic phosphoroclasm, formate dismutation and the fact that the hydrogen sulphate reaction can be stimulated by organic intermediates (see Grossman & Postgate, 1955). Hydrogen metabolism is mediated by reversible hydrogenases which are present in most strains of *Desulfovibrio*. Hydrogen uptake provided a valuable tool in early research on their metabolism because the stoichiometry and kinetics of substrate reductions could be studied by H_2 uptake experiments (e.g. Postgate, 1951a, b). Later the Knallgass reaction, a reaction between H_2 and O_2, mediated by cytochrome c_3 and resulting from the autoxidizibility of that pigment, was shown (see p. 74). Adams et al. (1951) reported 'hydrogenase-free' strains isolated from a sulphur-bearing lake in Cyrenaica (see p. 139) distinguished by failing to grow 'autotrophically' and being unable to reduce sulphate in hydrogen. The strains were lost before work on them was completed, and it is not known whether they would have shown hydrogenase activity in the presence of dyestuffs. One strain acquired the ability to reduce sulphate in H_2 after several subcultures in the presence of H_2.

The hydrogenase of *Desulfovibrio* can be exceptionally active, even in crude extracts and cell suspensions. The molar specific rate of reduction of methylene blue by cell suspensions can be as much as 20 times that of sulphate; hence reduction of this dye provides a more sensitive index of the presence of hydrogenase than does sulphate reduction. Methylene blue, however, inhibits hydrogenase in other organisms (Curtis & Ordal, 1954) as well as in *Desulfovibrio* (King & Winfield, 1955), an effect enhanced by NaCl (Littlewood & Postgate, 1956). Benzyl viologen or methyl violet are preferable test substrates despite the fact that methylene blue is usually the dye most rapidly reduced by intact cells.

The rates of reduction of other dyestuffs are not necessarily related to their standard redox potentials: in unpublished experiments by Miss J. Farmer and me, *Desulfovibrio vulgaris* reduced phenol-indol-dichlorophenol, Janus green or sodium indigodisulphate faster when the organisms had been treated with sufficient quaternary detergent (cetyltrimethylammonium bromide) to make them freely permeable to diffusible compounds. Hence the rate of dye reduction by intact cells is ordinarily limited by penetration of the dye to the enzyme. Methylene blue seems exceptional in this respect: its rate of reduction was the same whether the detergent was present or not. Bell, Le Gall & Peck (1974) showed that the soluble hydrogenase of *D. gigas* is readily released by osmotic shock and is probably localized in the periplasmic layer, a view supported by later studies on other strains of *Desulfovibrio* (Badziong & Thauer, 1980; Glick, Martin & Martin, 1980; Odom & Peck, 1981a). However, it is likely, though not rigidly proved, that hydrogenases exist both periplasmically and within the cytoplasm (see Peck & Le Gall, 1982).

Cell-free hydrogenase reduces dyes under H_2, and evolves H_2 from sodium dithionate + methyl or benzyl viologen, like any other reversible hydrogenase. Ferricyanide is reduced in H_2 by crude enzyme preparations (Sadana & Jagganathan, 1954), and Krasna & Rittenberg (1955) mentioned reduction of the nitroprusside ion. The selenite ion is also reduced. Callender & Roberts (1961), seeking analogies to nitrogen fixation, observed that azo-pyridine but not several other azo compounds were reduced by hydrogenase from *Desulfovibrio*. Krasna & Rittenberg (1954) demonstrated the exchange reaction H_2–D_2O and the ortho-para hydrogen conversion with *Desulfovibrio*. The enzyme seems to function in the dried state and will reversibly reduce a dry cytochrome (Kimura, Suzuki, Inokuchi & Yagi, 1979).

Hydrogenases have been extracted from *Desulfovibrio* by many workers and substantially purified by Sadana & Morey (1961), Yagi, Honya & Tamiya (1968), Haschke & Campbell (1971), Le Gall *et al.* (1971b), Yagi (1970), Hatchikian, Bruschi & Le Gall (1978) and others; see also Peck & Le Gall (1982). The consensus is that the soluble enzyme of *D. vulgaris* is a non-haem iron–sulphur protein, and molybdenum, once thought to be part of hydrogenases, is not present. Activity of *Desulfovibrio* hydrogenase differs according to species; a preparation obtained by van der Westen, Mayhew & Veeger (1978) from *D. vulgaris* had 12 iron atoms and 12 sulphur atoms per molecular weight of 50 000 and its specific activity, like that of Sadana & Morey's (1961) enzyme, was remarkably high. The affinity of *D. vulgaris* hydrogenase for H_2 (its K_m) is also very high (Kristjansson, Schönheit & Thauer, 1982). The high and efficient activity of the *D. vulgaris* enzyme has excited the attention of biotechnologists (see Postgate, 1982a). The hydrogenase of *D. gigas* contains nickel: it is a Ni-[4Fe-4S]$_3$ protein (Cammack et al., 1982; Le Gall et al., 1982; Moura et al., 1982). The Ni-atom gives an e.p.r.- (electron paramagnetic resonance) signal (about half of it is e.p.r.-active) with an E_m for its oxido-reduction about 280 mV more oxidizing than the hydrogen electrode. The enzyme from *D. gigas* appears to have two different subunits; it is of relatively low specific activity but Hatchikian & Monson (1980) reported that it is more active when immobilized. Many hydrogenases are unstable to oxygen and reports on the oxygen stability of *Desulfovibrio* hydrogenases are confusing: some degree of oxygen-damage can be shown in that activity of crude preparations is often stimulated by dithionite. Van der Westen, Mayhew & Veeger (1980) made the interesting observation that the soluble hydrogenase of *D. vulgaris* is sensitive to O_2 but, if the live cells are exposed to oxygen before the enzyme is extracted, an O_2-stable enzyme is formed.

The impression can readily be gained that a single organism contains multiple forms of hydrogenase (Ackrell, Asoto & Mower, 1966) but, if the example of *Clostridium pasteurianum* is any guide, such 'iso-enzymes' are operational artefacts (Nakos & Mortenson, 1971). Nevertheless, *D. vulgaris* appears to possess at least two molecular species of hydrogenase, one soluble and the other particulate (Yagi, 1970; Tsuji & Yagi, 1980; Martin, Glick & Martin, 1980). Growth conditions influence hydrogenase synthesis (Martin, Glick & Martin, 1980; Tsuji & Yagi, 1980) and, in a lactate–peptone medium, Tsuji & Yagi (1980) observed a burst of H_2 evolution at an early stage of growth which ceased as sufficient soluble hydrogenase was synthesized to 'trap' the H_2. They regarded the burst of H_2 evolution as arising from lactate in an initial ATP-generating process which primed the cell for the ATP-consuming reaction of sulphate reduction. Their experiments support the view that free H_2 generated metabolically from the carbon source is a normal part of the energy-yielding processes associated with sulphate reduction and point to roles of hydrogenase in both the formation and re-cycling of this H_2 (see also p. 95).

An important feature of the soluble hydrogenases of *Desulfovibrio* is their periplasmic location, which is probably essential to their role in energy generation (see Thauer & Badziong, 1981, also 'Energy metabolism' below, p. 91). This location also facilitates H_2 uptake from the external environment. *Desulfovibrio* frequently populates environments (muds, compost heaps, fermenting sludges) in which other anaerobes generate H_2; the possession of a peripheral hydrogenase would assist uptake of that H_2 as it is formed and its use by *Desulfovibrio* as an energy source (see also Chapter 7).

The hydrogenase of *Desulfotomaculum ruminis* is particulate (Buller & Akagi, 1964). Hydrogenase is not universal in this genus and it seems to be rare, though not entirely absent, among the slow-growing, acetate-utilizing types of sulphate-reducing bacteria.

Electron transport

The independent discovery by Ishimoto *et al.* (1954*a*) and Postgate[1] (1954) of a cytochrome in *Desulfovibrio* was historically an event of major importance, since proteins of this kind were thought to be restricted exclusively to aerobes. In retrospect, of course, the presence of cytochromes is entirely consistent with the oxidative character of their metabolism. The cytochrome of these bacteria was called cytochrome c_3[2] by Postgate (1956*a*) since it belonged to the '*c*' group, being relatively stable to heat, not easily degraded to produce a peptide-free haem group and its principal visible absorption band (its α-band) was at 553 nm, closer to the 550 nm of other cytochromes *c* than to the 560-nm range of cytochromes *b* (see Kamen & Horio, 1970). 'Cytochrome c_3' was purified and characterized by both the British and the Japanese research workers; the material was distinctive in having the lowest standard redox potential (around -0.2 V at pH 7) of any known cytochrome, as well as an iron content corresponding to two haem groups in a molecular weight of 12 000. It was autoxidizible: the reduced form became spontaneously oxidized in air. It

---

[1] Perhaps it is permissible to record that cytochrome c_3 was first seen, to my knowledge, by Dr June Lascelles (then at Oxford University) to whom I had sent a suspension of *Desulfovibrio vulgaris* strain Hildenborough early in 1952. Her telephoned accusation that an impure culture had been sent happily proved false!

[2] Cytochromes *c* and c_1 were present in muscle and c_2 had been adopted for a *c*-type cytochrome in photosynthetic bacteria. Cytochromes c_4 and c_5 were named in *Azotobacter vinelandii* shortly afterwards; thereafter a more rational nomenclature was adopted for cytochromes, according to which the original cytochrome c_3 ought to be called *Desulfovibrio desulfuricans* cytochrome 553 (the number refers to its principal visible absorption band). Unfortunately *D. desulfuricans* contains at least three cytochromes which would rate this designation. A proposal to apply the term c_3 to all bacterial low-potential cytochromes *c* (Meyer, Bartsch & Kamen, 1971) has not been widely accepted.

Table 5. *Some properties of the principal cytochromes of* Desulfovibrio

	Name[a]		
	c_3	c_{553}	cc_3
Molecular weight	13 000	6500 to 9000	~14 000
Isoelectric pH	10.5[b]	below 7	below 7
Absorption peak (reduced, nm)			
α	553	552.8	552.2
β	522	529	524
γ	419	414	420
Haem groups/molecule	4	1	4 to 8
E_0' (pH 7, mV)	-205 to -260	0 to -100	below -100

[a]Yagi's (1969) cytochrome has a molecular weight of 60 000 but is otherwise not characterized. Further data concerning the cytochromes listed were given by Ambler, Bruschi & Le Gall (1971). Evidence b- and d-type cytochromes in *Desulfovibrio* was presented by Jones (1971a).
[b]Cytochrome c_3 from *D. gigas* is isoelectric at pH 5.2.

is now clear that the material studied was contaminated with denatured material, that its haem content was even higher than was then suspected and that it was probably a mixture of distinct cytochromes all with rather similar absorption spectra (see Table 5); therefore a detailed account of its properties would not be useful. Nevertheless its involvement in the reduction of a variety of substrates was unequivocally demonstrated, using bacterial preparations made permeable by the detergent cetyltrimethylammonium bromide and also using crude cell-free extracts. With gaseous H_2 as reductant, 'cytochrome c_3' accelerated reduction of thiosulphate, tetrathionate or dithionite ions and its effect on sulphite reduction was marginal but real. Attempts to detect an effect on sulphate reduction were not successful because no active extracts able to reduce sulphate were available: the need for ATP (below) was not then recognized. Reduced 'cytochrome c_3' reacted spontaneously with O_2 and this fact led to the discovery of an unexpected reaction in *Desulfovibrio*: the bacteria synthesized water from a mixture of $H_2 + O_2$, a reaction which required 'cytochrome c_3' and for which it functioned as 'oxidase'. At low O_2 tensions, organic substrates such as lactate or pyruvate could reduce O_2 (see p. 69); nitrite or hydroxylamine could also react chemically with reduced 'cytochrome c_3' and nitrite or hydroxylamine reductase activity could thus be simulated. These and other aspects of the early history of 'cytochrome c_3' arose from the work of Ishimoto's group, of Senez's group and of Postgate; they were reviewed with references by Postgate (1959a, 1965a). Since the material discussed was not homogeneous and, in particular, there are now known to be several cytochromes with α-bands at or near 553 nm, it will be more constructive to describe the present stage of knowledge than to deal with the cytochromes of these bacteria historically. This is a research area in which progress is rapid, so the following account will be brief. For recent news the reader should consult current literature and review articles. Background literature was cited in reviews by Le Gall & Postgate (1973), Le Gall, Dervartanian & Peck (1979) and Peck & Le Gall (1982); Table 5 gives the principal properties of the major cytochromes of *Desulfovibrio*. Cytochromes c have been detected in *Desulfobacter postgatei* (Thauer, 1982), *Desulfobulbus propionicus* (Widdel & Pfennig, 1982) and others of the new genera (see Table 1 and Widdel, 1980).

Cytochromes c_3

This name is reserved for cytochromes corresponding to the dominant proteins described by the earlier British and Japanese workers. Cytochrome c_3 is the principal cytochrome of *Desulfovibrio vulgaris* and has four haem groups per molecule; its amino acid sequence has sufficient attachment sites (cysteine residues separated by two or four other amino acid residues) for four haem groups (Ambler, 1968). In contrast with mitochondrial and some other cytochromes, in which positions 5 and 6 of the haem are ligated to histidine and methionine residues, cytochromes c_3 are ligated to two histidine residues. This feature has been proposed as definitive for cytochromes c_3 (Probst, Bruschi, Pfennig & Le Gall, 1977). They are present in all species of *Desulfovibrio*. Those from *D. desulfuricans* and *D. salexigens* are electrophoretically different from each other and from that of *D. vulgaris*; they are also immunologically distinct according to the precipitin reaction (Drucker & Campbell, 1969). The cytochrome c_3 of *D. africanus* resembles that of *D. vulgaris* (Singleton, Campbell & Hawkridge, 1979). All four proteins are strongly basic and of low standard redox potential (below 0.2 mV); the actual value of the E_m is influenced by the method of measurement and is probably the average of values which differ as between the four individual haem groups (see Van Leeuwen *et al.*, 1982). The highly basic character of these cytochromes (isoelectric point about pH 10) distinguishes them from the cytochrome c_3 of *D. gigas*, which is isoelectric at pH 5.2; it, too, is a low-potential, tetra-haem protein with a distinctive amino acid sequence. The four haems of cytochromes c_3 appear to be close together in the molecule (Dobson et al., 1974).

Le Gall, Mazza & Dragoni (1965) extracted cytochromes c_3 from *D. gigas* simply by shaking the cells in water, which suggests that cytochromes c_3 probably exist in the living bacteria largely as soluble proteins in the periplasmic layer (see also Badziong & Thauer, 1980; Odom & Peck, 1981a; Peck & Le Gall, 1982). However, the apparent solubility is much affected by the method of extraction and the impression that cytochrome c_3 can be membrane-bound is readily formed. While a proportion of this protein may be bound *in vivo*, Peck & Le Gall (1982) took the view that cytoplasmic or bound cytochrome cc_3 (below) could confuse the issue.

The structures and sequences of some cytochromes c_3 and some other cytochromes from *Desulfovibrio* are known (see Ambler, Bruschi & Le Gall, 1971; Le Gall & Postgate, 1973; Bruschi, 1981; Shinkai, Hase, Yagi & Matsubara, 1980 and Haser et al., 1979 for references). Some cytochromes c_3 have been crystallized: that of *D. vulgaris* strain Miyazaki crystallizes in two forms (Bando et al., 1979). The 3-dimensional structure of cytochrome c_3 from *D. vulgaris* Miyazaki is now known (Higuchi et al., 1981).

Functionally, cytochrome c_3 seems to be an electron carrier in the reduction of sulphite, thiosulphate and other substrates. It is necessary for the reduction of flavodoxin, ferredoxin and rubredoxin (Yagi et al., 1968; Bell et al., 1978), and Xavier et al. (1979) have evidence from electron paramagnetic resonance data that the cytochrome c_3 of *D. gigas* interacts specifically with the electron carrier ferredoxin II (below). There is some evidence (Fauque, Barton & Le Gall, 1980) that colloidal sulphur can oxidize reduced cytochrome c_3 and that, in some strains of *Desulfovibrio*, cytochrome c_3 can act as a sulphur reductase allowing energy generation and growth. Reduced cytochromes c_3 are oxidized non-enzymically by nitrite, hydroxylamine and oxygen but these oxidations are not exploited physiologically (see Postgate, 1961). The electron transport function of cytochrome c_3 must obviously involve its haem groups; whether the possession of four such groups enables the protein to transfer four electrons at a time is uncertain, but ability to transfer more than one is indicated by cyclic voltammetry (Bianco, Fauque & Haladjian, 1979) as well as by the existence of four partially reduced states, deduced from electron paramagnetic resonance spectra (Le Gall, Bruschi-Heriaud & DerVartanian, 1971a; DerVartanian & Le Gall, 1971; DerVartanian, Xavier & Le Gall, 1978) and nuclear magnetic resonance (McDonald, Phillips & Le Gall, 1974). Yagi, Inokuchi & Kimura (1983) provided a useful if brief review of the properties of cytochromes c_3.

Electron transport 77

Cytochromes cc_3

Cytochromes have been found in *Desulfovibrio vulgaris, D. gigas* (Bruschi, Le Gall, Hatchikian & Dubordieu, 1969) and *D. desulfuricans* strain Norway (Guerlesquin, Bovier-Lapierre & Bruschi, 1982) which resemble cytochrome c_3 in that they have an α-band at 553 nm and a low potential. However, their molecular weights lie around 26 000. Their distinctive amino acid composition indicates that they are not simply aggregates of the classical cytochrome c_3, nor compounds of cytochrome c_3+cytochromec_{553}. They are probably most properly classified as higher molecular weight members of a cytochrome c_3 group of haemoproteins, a nomenclature adopted by Peck & Le Gall (1982). For convenience of reference I have retained the older nomenclature. Hatchikian, Le Gall, Bruschi & Dubordieu (1972) presented evidence that the cytochrome cc_3 from *D. gigas* is specifically involved in the reduction of thiosulphate.

Cytochromes c_{553}

These haemoproteins are present in strains of *Desulfovibrio vulgaris, D. desulfuricans, D. salexigens* but not *D. gigas* (Le Gall & Bruschi-Heriaud, 1968; Yagi, 1969; Bruschi, Le Gall & Dus, 1970; Peck & Le Gall, 1982) and have very similar absorption spectra. Their standard potential is more oxidizing (near -0.05 V) and their molecular weights lie between 8000 and 9000; amino acid analyses indicate that they are not degradation products of cytochrome c_3. They contain only one haem group/molecule. Yagi (1979) purified this protein from *D. vulgaris* strain Miyazaki and recorded a lower standard potential (-0.25 V) and what is probably the lowest molecular weight of any cytochrome: 8000. Yagi (1979) claimed that cytochrome c_{553} is a specific electron acceptor from formate, in conflict with an earlier report by DerVartanian & Le Gall (1974); he and his colleagues also reported that it accepts electrons from D-lactate (Ogata, Arihara & Yagi, 1981). Peck & Le Gall (1982) listed cytochrome c_{553} as a periplasmically-located protein.

Large cytochromes

Yagi (1969) mentioned briefly the presence of a large cytochrome with an α-band at 553 nm in his strain of *Desulfovibrio vulgaris*; its molecular weight was 60 000 (see Yagi, 1979).

78   *Metabolism*

Cytochrome b

Jones (1971*a*) reported a *b*-type cytochrome in *Desulfovibrio africanus* and one has also been detected in other species of *Desulfovibrio* (Hatchikian & Le Gall, 1972; Jones, 1972; Badziong & Thauer, 1980). This class of cytochrome is the dominant type in the genus *Desulfotomaculum* and has been detected in *Desulfobulbus propionicus* (Widdel & Pfennig, 1982) and *Desulfobacter postgatei* (Thauer, 1982) as well as in other representatives of the 'new' genera (see Widdel, 1980 and Table 1).

Cytochrome d

A protein of this class was also reported in *Desulfovibrio africanus* by Jones (1971*a*).

Cytochrome c_7

Desulfotomaculum acetoxidans has a *c*-type cytochrome with an α-band at 551 nm and a low E'_0 which is possbly a tri-haem protein (Probst *et al.*, 1977). It is similar to one from the mixed culture *Chloropseudomonas ethylica* (see p. 115) which was called c_7.

The cytochrome pattern in sulphate-reducing bacteria is thus much more complex than at first thought and it seems almost an act of perversity that *Desulfovibrio* should possess three or perhaps four cytochromes of very similar optical properties. The rationalization of the situation owes much to the impressive work of Dr Jean Le Gall and his associates; much more is known of the composition and character of these proteins than has been presented here.

In addition to the cytochromes there exist non-haem iron electron transport proteins resembling those characteristic of other obligate anaerobes.

Ferredoxins

These proteins are present in *Desulfovibrio, Desulfotomaculum* and probably other sulphate reducers, since they are found in most anaerobes. They are soluble proteins but located in the cytoplasm. One from *Desulfovibrio vulgaris* has been isolated and crystallized; its most distinctive feature is that it possesses four iron atoms per molecular weight of about 6000, in contrast with the eight-iron atoms of clostridial ferredoxin or the two-iron atom per molecule of plant ferredoxins (Zubieta, Mason & Postgate, 1973). Since the demonstration by Sieker, Adman & Jensen (1971) of two four-iron clusters in an eight-iron ferredoxin from a micrococcus (see Fig. 11), it is tempting to suggest that this *D. vulgaris* ferredoxin represents a naturally-occurring half molecule. *D. africanus* possesses three distinct ferredoxins, all of which are four-iron proteins (Hatchikian, Jones & Bruschi, 1979) but the third of which lacks the conventional arrangement of cysteinyl residues to bind its 4Fe–4S cluster (Bruschi & Hatchikian, 1982). *D. desulfuricans* strain Norway also possesses two ferredoxins (Guerlesquin, Moura & Cammack, 1982).

The situation in *Desulfovibrio gigas* is particularly interesting. Two ferredoxins exist, apparently as polymeric forms from a basic subunit of

Fig. 11. The 4Fe—4S cluster of ferredoxins. The sketch is based on the structure of that of the ferredoxin of *Peptococcus (Micrococcus) aerogenes* as demonstrated by Sieker, Adman & Jensen (1971) using X-ray crystallography. Two such clusters are present in that protein. Covalent bonding to the peptide chain takes place via cysteine residues symbolized by the S atoms drawn external to the 4Fe—4S cube.

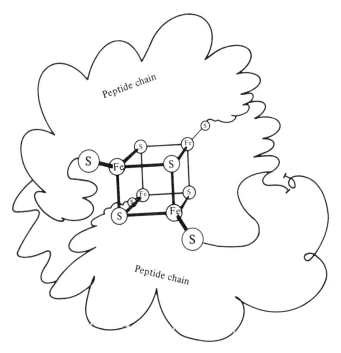

6000 daltons; one (Fd I) of 18 000 D and the other (Fd II) of 24 000 D (Bruschi et al., 1976). The amino acid sequence is known (see Bruschi, 1979). They have similar enzyme activities but electron paramagnetic resonance studies (Cammack et al., 1977) supported by proton magnetic resonance studies (Moura et al., 1977a) indicate that the properties of the iron–sulphur centres differ as between the polymers. A third ferredoxin was earlier thought to be present but proved to be a mixture of Fd I and Fd II. The remarkable feature of these ferredoxins is that Fd II contains a three-iron cluster (Huynh et al., 1980) and that Fd I contains a mixture of three-iron and four-iron clusters (Johnson et al., 1981). The two ferredoxins differ in their physiological activity, Fd I being more active in the pyruvic phosphoroclasm, Fd II being more active in sulphite reduction (Moura et al., 1978). Both can be reduced by hydrogen + hydrogenase but cytochrome c_3 is needed as a carrier. The three-iron cluster of Fd II can be converted into a four-iron cluster *in vitro* (see Kent et al., 1982) which suggests that comparable transformations of the prosthetic site may occur *in vivo*.

Rubredoxins

These proteins, pink when oxidized, colourless when reduced, contain one iron atom per molecular weight of about 6000 and no labile sulphide, and they resemble the rubredoxins of other bacteria. Rubredoxins have been isolated and purified from at least five species of *Desulfovibrio*: *D. vulgaris, D. desulfuricans, D. gigas, D. africanus* and *D. salexigens*. A rubredoxin has also been isolated from the sulphur-reducing *Desulfuromonas acetoxidans* (Moura et al., 1979). The rubredoxins seem to be rather similar in amino acid composition and redox properties from all these sources (see Moura et al., 1979; Hatchikian, Jones & Bruschi, 1979; Moura, Moura, Bruschi & Le Gall, 1980); their mid-point potentials are in the range −46 to +6 mV. Data on the X-ray structure of two crystalline desulfovibrio rubredoxins have been published (Pierrot et al., 1976; Adman et al., 1977). A protein which may be related to the rubredoxins, called 'desulforedoxin', has been isolated from *Desulfovibrio gigas* (Moura et al., 1977b); it contains two iron atoms per molecular weight of 7900 and no labile sulphur though it contains the eight cysteine residues necessary for the rubredoxin-like iron centres. Its sequence is known: it comprises two 36-fold polypeptide chains (Bruschi et al., 1979) and details of its Mössbauer and e.p.r. spectra are available (Moura, I. et al., 1980). Like rubredoxin, its pink colour is bleached by dithionite.

No function has been assigned to these proteins and my assumption that they have an electron-transport role is speculative.

Flavodoxins

These have been isolated from *Desulfovibrio gigas, D. vulgaris* and *D. salexigens*; they are soluble, cytoplasmic proteins. They are rather alike in spectra and amino acid sequence (Dubordieu & Fox, 1977; Moura, Moura, Bruschi & Le Gall, 1980). They are flavoproteins of molecular weight about 16 000 with flavin adenine mononucleotide as the prosthetic group; they form blue semiquinones which may be the cause of the blue, autoxidizible tint occasionally seen in cultures of *D. salexigens*. The X-ray structure of *D. gigas* flavodoxin has been reported in considerable detail (Watenpaugh *et al.*, 1972). Flavodoxins will substitute for ferredoxins in the pyruvic phosphoroclasm and the flavodoxin of *D.salexigens* is active in its sulphite reductase system (Moura, Moura, Bruschi & Le Gall, 1980). Peck & Le Gall (1982) mentioned evidence that the flavodoxin of *D. gigas* undergoes reversible phosphorylation which alters its specificity: only the phosphorylated form intervenes in sulphite reduction but both forms can participate in pyruvate breakdown.

Menaquinones

The menaquinone MK6 has been detected in *Desulfovibrio gigas, D. desulfuricans* and *D. vulgaris* (Morton, Matschiner & Peck, 1970; Maroc, Azerad, Kamen & Le Gall, 1970; Badziong & Thauer, 1980); it is not otherwise known in anaerobes though it occurs in aerobes.

There are today too many oxido-reducible factors in sulphate-reducing bacteria for a rational picture of their electron transport to be assembled. An impressive but incomplete synthetic survey was made by Peck & Le Gall (1982). In *Desulfovibrio* it seems reasonable to suppose that the cytochromes are directly involved in transport of electrons to reducible substrates such as thiosulphate, tetrathionate and sulphite and, since sulphite seems to form part of the normal pathway of sulphate reduction (see the next section), they would thus be of critical importance in normal sulphate reduction. However, cytochrome $_{553}$ seems to be specifically involved as electron acceptor from formate and D-lactate, though cytochrome c_3 can also accept electrons from formate (Ambler *et al.*, 1971). Moreover, there is now good evidence that sulphite reduction requires ferredoxin as well as a cytochrome (perhaps this is why the effects of c_3 on sulphite reduction were relatively slight in the hands of the early workers), and flavodoxin, but not rubredoxin, can substitute for ferredoxin. Cytochrome cc_3 is particularly reactive in the reduction of thiosulphate, and the cytochrome *b* of *Desulfovibrio gigas* seems to be specifically concerned in the reduction of fumarate. Ferredoxin is involved in pyruvic phosphoroclasm and reductive carboxylation of acetyl phosphate in both *Desulfovibrio* and *Desulfotomaculum*; cytochrome c_3 appears to augment its effect in the former genus. An NADH-rubredoxin oxido-reductase and an NAD(P)H-menadione oxido-reductase exist in *Desulfovibrio gigas*, but otherwise the roles of rubredoxin and menadione MK-6 are obscure.

A problem of principle in assigning roles to electron-transport factors of low potential is that artificial effects, such as the reaction of cytochrome c_3 with O_2 or nitrite, can occur and may obscure the position. Does the fact that cytochrome c_3 augments the ferredoxin-dependent pyruvic phosphoroclasm in *Desulfovibrio* imply that it has a physiological function in this reaction? Not necessarily. Moreover, in most of their reactions, these carriers can be replaced by a low-potential redox dye such as methyl viologen, so how is one to be certain that a carrier such as flavodoxin, in augmenting thiosulphate reduction, is not having an artificial, viologen-like effect? Specificity for a given reaction, such as those of cytochrome *b* towards fumarate reduction or cc_3 towards thiosulphate reduction, provides more convincing evidence for a physiological role but, as so often with comparable low-potential carriers, evidence of such specificity is usually lacking in sulphate-reducing bacteria. One instance of electron transfer has been studied with pure proteins: cytochrome c_3 of *Desulfovibrio gigas* donates an electron to the large polymeric ferredoxin, but the process seems to involve little site specificity (Moura *et al.*, 1977c). For present purposes one can only record that the position is confused and point out that the presence of substances characteristic of both aerobes (cytochromes, menadione) and of anaerobes (ferredoxins, rubredoxins and flavodoxins) underlines the curiously ambivalent physiological status of this group of microbes.

Dissimilatory sulphur metabolism

Where aerobes reduce oxygen to water, the sulphate-reducing bacteria reduce sulphate to water plus sulphide. The manner in which they do this, and the nature of the enzymes involved, has interested scientists since the earliest days of research in this area – the classical thesis of Baars (1930) includes a hypothetical chemical pathway from sulphate to sulphide. Elucidation of the pathway has been complicated by the fact that most of the potential intermediates are unstable and interconvertible, so the risk of experimental artefact is considerable. However, oblique kinetic evidence that the sulphite ion is an intermediate, but that thiosulphate and tetrathionate are not, appeared in early biochemical publications (Postgate, 1951b; Koyama, Tamiya, Ishimoto & Nagai, 1954). Such early work was summarized by Postgate (1959a); though more recent findings have superseded most of the early conclusions, two important pieces of information were obtained.

To Millet (1955) goes the credit of establishing that sulphite, or some molecule biochemically equivalent thereto, is an intermediate in normal sulphate reduction, since she isolated radioactive sulphite from a population in the process of reducing radioactive sulphate.

A second interesting finding concerned the structural specificity of the sulphate-reducing system: the sulphate ion has several structural analogues (Fig. 12) and, of these, the selenate and monofluorophosphate ions

Fig. 12. Some structural analogues of sulphate. Only those with two charges affect sulphate reduction by *Desulfovibrio vulgaris*. Of those, the effects of selenate and monofluorophosphate are specifically on the sulphate-reducing enzyme system, but chromate, though a useful inhibitor in practice, is not specific: it exerts its effect through its oxidizing properties (after Postgate, 1952b).

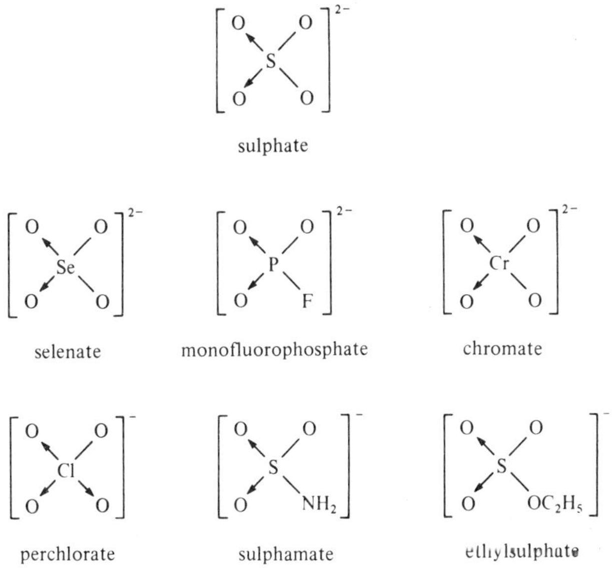

sulphate

selenate   monofluorophosphate   chromate

perchlorate   sulphamate   ethylsulphate

proved to be powerful and specific competitive inhibitors of sulphate reduction (Postgate, 1949, 1952b), though not of the reduction of ions such as sulphite or thiosulphate. The interest of selenate inhibition was that the inhibition index – the molar ratio of inhibitor of substrate giving half-maximum reaction rate – was about 0.025, indicating that the enzyme system had some 40-fold greater affinity for the inhibitor, selenate, than for its natural substrate. Furusaka (1961) presented evidence that selenate acts against the accumulation of sulphate in the organism, rather than on its actual reduction. The interest of monofluorophosphate inhibition lay in the fact that its structural analogy to sulphate was not immediately obvious: the P–F bond is very similar in mass and electron distribution to the S–O bond. Other sulphate analogues were not specific inhibitors, but the chromate ion proved to be a very powerful non-specific growth inhibitor of sulphate-reducing bacteria, one which has proved useful in practice.

The molybdate ion, which is formally a structural analogue of sulphate, is also an inhibitor of sulphate reduction; like chromate, but unlike selenate, it appears to inhibit *Desulfovibrio* by forming an unstable analogue of active sulphate (below) and thus depleting the cell of ATP (Taylor & Oremland, 1979). The effect of molybdate on sulphate-reducing bacteria is relatively specific and it has proved to be a useful inhibitor of sulphate reduction in ecological studies (see Chapter 7).

Cells of *Desulfovibrio* are impermeable to sulphate on a gross scale[1] (Littlewood & Postgate, 1957a) but entry of sulphate does take place because free sulphate can be detected within cells which are reducing radioactive sulphate (Furusaka, 1961).

The process of sulphate reduction is now known to involve consumption of ATP, a fact established by Peck (1959) and Ishimoto (1959); thereafter a mechanism which may be cyclic leads to sulphide formation. The elucidation of these processes has already taken more than a decade and the details of the sulphide-generating system are still uncertain. Some discussion of alternative schemes, and references to background literature, may be found in the reviews by Le Gall & Postgate (1973), Akagi (1981) and Peck & Le Gall (1982). Some of the features of the sulphate reduction pathway follow.

---

[1] In unpublished experiments I confirmed the gross impermeability to sulphate using the 'optical effect' (Mager, Kuczynski, Schatzberg & Avi-dor, 1956) of sodium sulphate and chloride on *D. vulgaris* suspensions; glycerol produced no optical effect, indicating the absence of a permeability barrier to this solute.

Activation of sulphate. The first step of dissimilatory sulphate reduction is the activation of the sulphate ion: its conversion to an adenosine phosphosulphate (APS for short). APS has the structure below, in which two of the phosphate residues of ATP are replaced by a sulphate group.

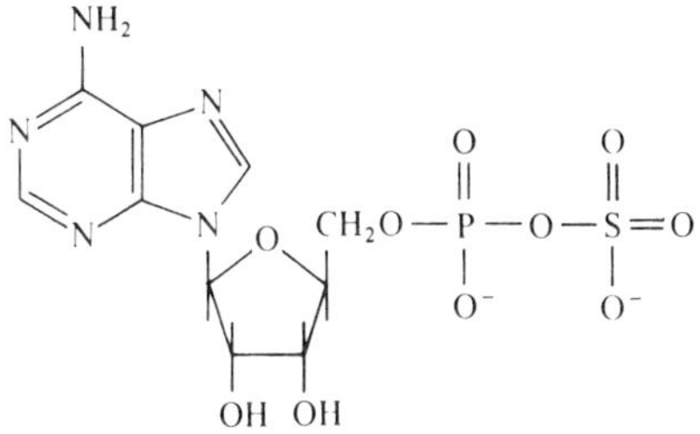

It is formed by ATP-sulfurylase (otherwise called sulphate adenylyl transferase) from ATP and sulphate by the reaction which also yields pyrophosphate (PP):

$$ATP + SO_4^{2-} \rightarrow APS + PP$$

The enzyme has been extracted and partly purified from *Desulfovibrio* (Baliga, Vartak & Jagannathan, 1961) and from *Desulfotomaculum nigrificans* (Akagi & Campbell, 1962). The enzyme from the thermophile shows increased heat stability. APS is also formed during assimilatory sulphate reduction, but a doubly phosphorylated 'active sulphate' is then formed, called phospho-adenosinephosphosulphate (PAPS), before reduction occurs. In dissimilatory sulphate reduction only APS is formed. The equilibrium constant of ATP-sulfurylase is not favourable to APS formation, so the reaction is 'pulled' to the right by an inorganic pyrophosphatase, which hydrolyses the pyrophosphate. This enzyme, in sulphate-reducing bacteria, has the property of being activated by reducing agents, and the organism inactivates it when the environment is aerobic. Thus, according to Ware & Postgate (1971), the organism can conserve ATP (preventing its conversion to APS) in conditions in which, for other reasons, growth cannot occur. ATP-sulfurylases from different strains show considerable differences (Skyring & Trudinger, 1973).

Reduction of adenosine phosphosulphate. The enzyme responsible for the reduction of APS is called APS-reductase (otherwise adenylyl sulphate reductase). It reduces APS to sulphite + AMP:

$$APS \underset{-2e}{\overset{+2e}{\rightleftharpoons}} AMP + SO_3^{2-}$$

As indicated above, it is reversible, and it is found working in the reverse direction in thiobacilli, playing a part in the oxidation of sulphur to sulphate; it has also been found in aerobic, photosynthetic sulphide-oxidizing bacteria and seems to be a definitive enzyme of microbes which conduct dissmilatory sulphur metabolism (see Peck, 1968). In laboratory tests, reduced methyl viologen is the customary electron donor for APS reduction, but cytochrome c_3 will act in place of the viologen and both ferredoxin and flavodoxin are stimulants. The enzyme has been purified from *Desulfovibrio vulgaris*; it is located in the cytoplasm and appears to be a trimeric flavoprotein (mol. wt $\sim$ 218 000) which contains one flavin –adenine dinucleotide group and Fe–S clusters (12 atoms of each/mole). There is evidence that the sulphur is actually bound to the flavin group as a sulphite adduct during the enzyme's action and that the non-haem iron group is the site at which electrons are accepted from whatever is the natural electron donor (see Peck & Le Gall, 1982).

Reduction of sulphite. The sulphite reductase system – the enzyme complex which yields sulphide from sulphite – is often associated with subcellular particles but soluble enzymes with differing properties have been obtained from *Desulfovibrio* and *Desulfotomaculum* species. In *Desulfovibrio* the bulk of the sulphite reductase is cytoplasmic. Both assimilatory and dissimilatory sulphate reductases may be present (e.g. Drake & Akagi, 1976). Le Gall & Postgate (1973) listed six kinds of sulphite reductase isolated from various sulphate-reducing bacteria and the position has, if anything, become more complex since then (see Peck & Le Gall, 1982; Akagi, 1981). Some reductase preparations produced sulphide from sulphite, others produced trithionate. The products are often different according to the electron carrier present: with methyl viologen, sulphide is often formed whereas a natural transporter might yield largely trithionate. The concentration of electron donor (e.g. pyruvate) may influence the products (Drake & Akagi, 1977) as does the pH and the concentration of electron carrier (Jones & Skyring, 1975). Thiosulphate can also be formed, and is reduced by extracts of *Desulfovibrio*; so are the tetrathionate and dithionite ions. Crude extracts may form exclusively sulphide but soluble preparations (e.g. from *D. vulgaris*) give thiosulphate and trithionate, the ability to give sulphide being restored by adding back a membrane fraction (Drake & Akagi, 1978). The ability of enzyme preparations to produce, and sometimes (but not always) to utilize, such polythio-ions led to proposals of cyclic reaction mechanisms in which they appear as intermediates in sulphite reduction. Le Gall & Postgate (1973) quoted examples proposed by Kobayashi, Tachibana & Ishimoto (1969) and by Akagi and his colleagues (Suh, Nakatsukasa & Akagi, 1968; Findley & Akagi, 1969, 1970; Suh & Akagi, 1969).

Fig. 13 presents a hybrid mechanism of a cyclic type. Such cyclic mechanisms have, however, been severely criticized by Chambers & Trudinger (1975) who took the view from labelling experiments with ^{35}S that trithionate and thiosulphate are not normal intermediates in sulphite reduction and that a direct reduction to sulphide takes place: they maintained that trithionate and thiosulphate are by-products. This view is plausible, because thiosulphate and polythionates are certainly formed spontane-

Fig. 13. A possible cyclic pathway for dissimilatory sulphate reduction. The sulphate ion, outside the cell, is accumulated by a process which selenate inhibits competitively. Once inside it reacts with ATP to form adenosine phosphosulphate (APS) plus pyrophosphate (PP), a reaction which only proceeds to the right because the pyrophosphate is removed as inorganic phosphate (P). APS is reduced to sulphite + AMP. Sulphite dehydrates to metabisulphite which is reduced *via* intermediates (the dithionite ion, $S_2O_4^{2-}$, is wholly speculative) to give trithionate ($S_3O_6^{2-}$). This is reductively split to give thiosulphate and to regenerate some sulphite; the thiosulphate is reduced to give sulphide and more sulphite. Enzyme preparations capable of forming the unbracketed components from the unbracketed precursors have been isolated from *Desulfovibrio* species and some from *Desulfotomaculum*; for comment on the scheme, see text.

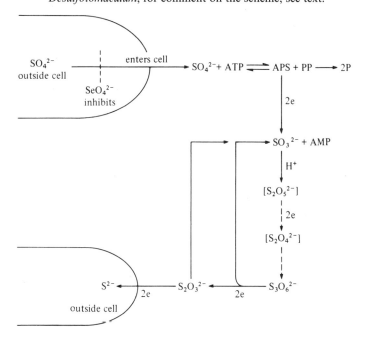

ously by the interaction of sulphide and sulphite. However, the question must be considered open at the time of writing for several reasons. One is that sirohydrochlorin, part of the prosthetic group of sulphite reductase (below), is easily removed from the enzyme during extraction and handling, and in the presence of iron it is capable of catalysing non-enzymic reduction of sulphite to sulphide (Seki & Ishimoto, 1979); in the absence of iron it catalyses reduction of dithionite (Skyring & Jones, 1977). Such non-enzymic processes might be complicating interpretation of enzyme studies. Another is that at least one strain of *Desulfovibrio* generates thiosulphate during normal sulphate reduction (Vainshtein *et al.*, 1980); a third reason is that it is difficult to account for the large amounts of trithionate-producing enzymes apparently present in *Desulfovibrio* entirely in terms of chemical artefacts. Akagi (1981) concluded that a cyclic pathway was highly probable but Peck & Le Gall (1982) took an opposing view.

Despite uncertainty about the mechanism of sulphite reduction, the study of this process revealed a function for an enigmatic green protein, desulfoviridin (earlier spelled 'desulphoviridin') which is abundant in *Desulfovibrio* (Postgate, 1956a). It proves to be the dissmilatory sulphite reductase. *In vitro* its principal product is trithionate (Lee & Peck, 1971; Kobayashi, Takahashi & Ishimoto, 1972; Seki, Kobayashi & Ishimoto, 1979) but sulphide is also formed (Skyring & Trudinger, 1972, 1973). Though earlier thought to be a porphyroprotein (Postgate, 1956a), desulfoviridin proves to be a haemoprotein which also possesses iron–sulphur clusters. Two chromatographically distinct forms with similar enzyme activities were isolated from *D. vulgaris* by Seki, Kobayashi & Ishimoto (1979). Desulfoviridin is located in the cytoplasm; it is an $\alpha_2\beta_2$ heteromeric tetramer, mol. wt. $\sim 200\,000$, with about four 4Fe–4S clusters and two haem groups per molecule. The haems are of a special class, the sirohaems, which are only otherwise found in the assimilatory sulphite reductases such as that of *Escherichia coli* (which also contains flavin) and in nitrite reductases. The sirohydrochlorin, from which the sirohaem is derived by insertion of iron, is very easily released from the protein: mild acid or heat treatment extracts it. Structurally it is related to the cobyrinic acid of vitamin B_{12}; sirohydrochlorin is a highly carboxylated,[1] water-soluble molecule which is unstable in light and may represent a rather primitive or ancestral type of porphyrin among living things. An account

---

[1] A wild but nearly correct guess by Postgate (1956a) was courteously applauded by Murphy & Siegel (1973) in their report of this novel haem.

of the sulphite reductases was presented by Siegel (1978); the structure of sirohydrochlorin is probably that below.

[Chemical structure of sirohydrochlorin showing a porphyrin-like macrocycle with substituents: CH₂CH₂COOH, CH₂COOH, CH₂CH₂COOH, CH₃, CH₂COOH, CH₂CH₂COOH, HOOC.CH₂CH₂, HOOC.CH₂, HOOC.CH₂]

Although redox changes would be expected during sulphite reduction, desulfoviridin preparations usually show no spectral changes on reduction with dithionite or oxidation with ferricyanide. Skyring & Jones (1976), however, obtained a species from *D. gigas* which did show such redox changes. Hall, Prince & Cammack (1979) and Liu, DerVartanian & Peck (1979) showed that the e.p.r. signals assignable to the sirohaem (observed in the oxidized form) became replaced by non-haem iron signals on reduction by methyl viologen; a free radical intermediate was transiently observed, but the extent of involvement of the haem and Fe-S groups in sulphite reduction is still not clear.

Desulfovibrio desulfuricans strain Norway 4 (Appendix 1) is one of the few strains of *Desulfovibrio* which lacks desulfoviridin; it is apparently replaced by a red protein termed (with more attention to rhyme than etymology) 'desulforubidin' by its discoverers (Lee, Yi, Le Gall & Peck, 1973). Its prosthetic group is also of the sirohaem class (Murphy *et al.*, 1973). The sulphite reductase of *Desulfotomaculum nigrificans*, termed P 582 (Trudinger, 1970) also has sirohaem as prosthetic group; both show e.p.r.–redox properties similar to those of desulfoviridin (Liu, DerVartanian & Peck, 1979).

Fractionation of sulphur isotopes. A phenomenon of importance for studying the geochemical activities of sulphate-reducing bacteria is the fractionation of sulphur isotopes that accompanies sulphate reduction. Natural sulphur consists of four stable isotopes, ^{32}S, ^{33}S, ^{34}S and ^{36}S, of which ^{32}S is the major component (about 95%) and ^{34}S is subsidiary (about 4.2%). During bacterial sulphate reduction the sulphate ions containing ^{32}S react slightly more frequently than those with ^{34}S. Thus the lighter sulphate becomes preferentially converted to sulphide and, while sulphate reduction is in progress, a mass-spectrographic analysis of the products shows that the sulphide is enriched in ^{32}S relative to geochemical sulphur and the residual sulphate tends to be enriched with ^{34}S. Isotopic fractionations of this kind occur during the biological transformations of carbon compounds also. In principle the mechanism is reasonably simple: the lighter mass and greater mobility of the ions containing the light isotope enable them to diffuse more readily to the site of sulphate reduction and thus to become reduced more easily. Formally, the process is analogous to the diffusion method used for the physical separation of isotopes.

Several studies of the conditions favouring isotope fractionation by these bacteria have been published (e.g. Thode, Kleerekoper & McElcheran, 1951; Jones & Starkey, 1957; Kaplan & Rittenberg, 1962, 1964a; McCready & Krouse, 1980). Temperature, sulphate concentration and growth rate of the organism all influence the extent of isotope fractionation. Accounts of the thermodynamics and kinetics of the process are available (Tudge & Thode, 1950; Harrison & Thode, 1958). An exhaustive study, which included other sulphur bacteria than sulphate reducers, was that performed by Kaplan & Rittenberg (1964a) which supported the general impression from earlier work that slow metabolism engendered by a low temperature favoured fractionation. Sulphite, which was metabolized faster than sulphate in all test conditions (but one), consistently gave lower fractionation figures than did sulphate. A curious feature of these studies is that the extents of fractionation obtained are often greater than those calculated for simple uni-directional reaction steps from substrate to sulphide, and the intervention of internal pools of sulphate, and of activation and reversible reactions, has been postulated (see Kaplan & Rittenberg, 1964a).

Reduction of other sulphur-containing substrates

Trithionate is reduced by *Desulfovibrio* suspensions to sulphide. *Thiosulphate* is also reduced by suspensions of most strains, but relatively slowly, and by their cell extracts. Extracts reduce thiosulphate to sulphite and sulphide (Ishimoto & Koyama, 1957), the sulphide originating from the outer (sulphone) sulphur atom according to tests with ^{35}S-labelled thiosulphate (Nakatsukasa & Akagi, 1969). Thiosulphate reductase is located in the cytoplasm but Peck & Le Gall (1982) reached the rather unsatisfying conclusion that it was not involved in energy generation. Yet *D vulgaris* can grow with thiosulphate as electron acceptor. *Tetrathionate* is reduced, more slowly than thiosulphate. *Dithionite*, though a strong reducing agent itself, is also reduced quite rapidly by cells and cell extracts. Postgate (1951*a, b*) concluded that *elemental sulphur* is not reduced but Biebl & Pfennig (1977) did not agree (though they published no $S^0 \to S^{2-}$ balances). Fauque, Barton & Le Gall (1980) regarded cytochrome c_3 as an effective reductase for colloidal sulphur in *D. gigas* and the Norway 4 strain of *D. desulfuricans*; the reaction could be conducted by membrane preparations and coupled, under H_2, to ATP synthesis. Several partially oxidized derivatives of sulphur were considered as substrates and rejected (see Postgate, 1959*a*). The possibility that disulphur monoxide (S_2O) may be involved was raised by Iverson (1967) but its characterization was not unequivocal.

Energy metabolism

Peck (1962) pointed out that a source of ATP other than substrate-level phosphorylation must exist in *Desulfovibrio*. His logic, which was compelling, amounted to this: since the reduction of sulphate consumes 1 molecule ATP per sulphate ion and yields 1 molecule of AMP, two molecules of acetyl phosphate would be needed to regenerate a molecule of ATP from that AMP. If the chemical equations for the oxidation of lactate to acetate using sulphate are written as in Fig. 14, it is seen that two lactate ions are needed to obtain those two acetyl phosphate ions. It is clear that the ATP economy balances exactly: one molecule of ATP is generated from lactate at the substrate level and one is consumed at the sulphate

Fig. 14. The ATP economy of bacterial sulphate reduction. The two molecules of lactate consumed provide just enough energy-rich phosphate, as acetyl phosphate, to regenerate one molecule of ATP from the one molecule of AMP formed (from ATP) when one sulphate ion is reduced to sulphide. Therefore further sources of ATP are necessary for growth with lactate (see text).

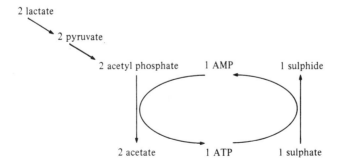

reduction level. It follows that the organisms could not possibly grow by these reactions unless they had another mechanism of phosphorylation. Yet they do grow, and the respiratory chain is the most probable alternative system for ATP generation.

Work with cell-free extracts of *Desulfovibrio* leaves little doubt that respiratory chain phosphorylation – generation of ATP during electron transport – does occur in these bacteria. Peck (1966) demonstrated ATP generation during the reduction of sulphite in H_2 by particles from *Desulfovibrio gigas*, a process now known to require ferredoxin, for which flavodoxin did not substitute (Barton & Peck, 1970). Fumarate can also serve as electron acceptor for respiratory chain phosphorylation (Barton *et al.*, 1970), providing further evidence for the special position of the fumarate–succinate couple in the metabolism of the bacteria mentioned earlier in this chapter.

Energy budgets can be deduced from growth yield experiments and a voluminous literature has grown up, based on the determinations of yield coefficients for bacteria grown with various carbon sources and the calculation therefrom of the ATP yields (Y_{ATP}) (see Stouthamer, 1977 for a guide to this topic). Such experiments present special difficulties when applied to sulphate-reducing bacteria for reasons presented by Postgate (1965a). These are the following.

1. Growth of sulphate-reducing bacteria is usually linear, not exponential. This phenomenon occurs because sulphide, although a product of metabolism, is also an inhibitor of growth: the higher the sulphide concentration, the longer the doubling time of the population. Thus energy is being expended in countering the inhibitory effect of sulphide and, in a batch culture, this energy loss interferes with the Y_{ATP} calculation because the so-called 'maintenance coefficient' increases as the culture grows.

2. If this problem is avoided by removal of sulphide, the pH shifts to an alkaline value, imposing another stress on the population which also requires 'maintenance' ATP.

3. Sulphide precipitates Fe, so that Fe, and not the carbon/energy source, may become the limiting substrate.

These considerations cause yield studies with sulphate-reducing bacteria, when performed on batch cultures, to be of rather variable scientific value and the precise experimental design should be borne in mind when assessing published data. Chemostat culture, in which parameters such as pH, sulphide concentration and limiting substrate can be controlled and maintained at known levels over many generations, is the technique of choice for such experiments although it, too, has a limitation: it is inappropriate for the determination of maximum growth rates. As an example of the problems that can arise, Magee, Ensley & Barton (1978) found that the yields of cells of *Desulfovibrio* with lactate were to all intents and purposes the same whether batch cultures were grown to stationary phase with sulphate ($Y_{lactate} = 6$ g/mole) or sulphite ($Y_{lactate} = 5.7$ /mole). Yet one would expect sulphite to give a greater yield because no ATP is expended in its activation and, indeed, the chemostat study of Hill, Drozd & Postgate (1972) had shown substantially different yields from lactate as between sulphate (6.3 g/mole) and sulphite (9.2 g/mole) as electron acceptors. One is tempted to the view that, since sulphate and sulphite would yield similar concentrations of sulphide in a batch culture, it was the sulphide and not the energy derived from the lactate that determined the yield in the batch cultures. Similar considerations probably account for the common observation (e.g. Traore *et al*., 1981) that pyruvate supports some 20% higher growth yields than does lactate in batch cultures: pyruvate generates half as much sulphide/mole as does lactate.

Science is frequently a matter of compromise, and disregard of these considerations has led to some important observations. Vosjan (1970) estimated the phosphorylation ratio, the ATP/electron transferred down the respiratory chain, at below 0.25. Traore *et al*. (1981) regarded the greater cell yield in pyruvate compared with lactate as due, not to lower sulphide formation, but to greater respiratory chain phosphorylation in the presence of the former substrate. Vosjan (1974) had shown that H_2 was evolved from pyruvate during sulphate reduction, the amount being influenced by the growth rate; Traore *et al*. (1981) found H_2 among the products of both lactate and pyruvate dissimilation during sulphate reduction. Indeed, there is now clear evidence that hydrogen is evolved, not only during the sulphate-free metabolism of pyruvate (p. 66), but also as a significant product of the metabolism of both pyruvate and lactate: as was discussed earlier (p. 72), Tsuji & Yagi (1980) showed that H_2 evolution from lactate by *D. vulgaris* was greatest at the early stages of growth. Obviously this reaction represents a net waste of ATP-generating potential and further complicates the interpretation of substrate yield experiments; in their experiments, removal of H_2 in a stream of argon lowered the cell yield.

The major criticism of batch culture experiments in this context is the pronounced inhibitory effect of sulphide. This criticism does not apply to the interesting experiments reviewed by Thauer & Badziong (1981, see also Badziong & Thauer, 1978) who studied batch cultures of *D. vulgaris* growing mixotrophically (at the expense of H_2, assimilating CO_2 + acetate in equimolar proportions – see p. 45). The pH values of their cultures were controlled and they were sparged with $H_2 + CO_2$ so, since the latter displaced H_2S, they grew exponentially to give a cell yield presumably uninfluenced by sulphide inhibition. Since hydrogen was the sole electron donor, all ATP would be generated by respiratory chain phosphorylation. By making certain assumptions concerning the Y_{ATP} (the yield of cell/-mole ATP) they calculated that the organism gained 1 mole ATP/mole sulphate reduced. They argued that sulphate reduction effectively consumed two ATP equivalents/sulphate ion reduced (one required to form APS, the other equivalent to the energy loss involved in hydrolysing pyrophosphate to 'pull' APS formation) and that an electron acceptor which did not require activation by ATP would allow a net yield of 3 mole ATP/mole acceptor. Sulphite was not experimentally exploitable but growth yields with thiosulphate, which is thermodynamically equivalent to sulphate, were approximately 3-fold greater. The strain studied by Thauer & Badziong (1981) appears to have been remarkably efficient in its coupling of energy generation from H_2 to biosynthesis.

Cell yields of desulfotomacula are always lower than yields of desulfovibrios in similar media and, although cytochromes are present, hydrogenase is often absent. Liu & Peck (1981) investigated this phenomenon and observed that the level of pyrophosphatase is remarkably low in *D. nigrificans*, *D. orientis* and *D. ruminis*. Since continuous removal of pyrophosphate is necessary for APS formation (p. 85), a paradox arose which was resolved by their discovery that these desulfotomacula contain the enzyme pyrophosphate: acetate phosphotransferase. This enzyme actually forms acetyl phosphate from acetate + pyrophosphate, thus providing an energy gain by a reaction which *Desulfovibrio* cannot exploit. Growth yield experiments suggested that this was the sole means of ATP generation for biosynthesis: that respiratory chain phosphorylation did not occur in this genus. If this view is correct the analogy between sulphate reduction and aerobic respiration mentioned earlier in this chapter (pp. 59–61) breaks down for this genus. Whether other sulphate-reducing bacteria share the ability of these *Desulfotomaculum* species to recover energy from pyrophosphate is not yet known, but the presence of an active pyrophosphatase in the acetate-utilizing *Desulfobacter postgatei* (Thauer, 1982) suggests that this species, at least, shares the *Desulfovibrio* type of energy-generating system.

Respiratory chain phosphorylation in aerobes is associated with vectorial proton and electron transfer (see, for example, Garland, 1977). A model for a chemiosmotic process for energy generation in sulphate-

Energy metabolism

reducing bacteria proposed by Wood (1978) suggested that periplasmic enzymes such as cytochrome c_3 and hydrogenase were irrelevant in this context, a view which was not accepted by Badziong & Thauer (1980), who preferred to regard the periplasmic enzymes as generating electrons for vectorial transfer across the cell membrane for cytoplasmic reduction of sulphate (or APS). Steenkamp & Peck (1981) studied the strain *D. desulfuricans* 27774 which is capable of nitrate reduction (below) and demonstrated a nitrite-dependent proton translocation out of the cell. Such a pattern would be consistent with the periplasmic location of the nitrite reductase and also fits in with current ideas on nitrate respiration in *E. coli* (see Jones, 1981). Kobayashi, Hasegawa, Takagi & Ishimoto (1982) have since demonstrated export of protons by *D. vulgaris* promoted by sulphite and augmented by H_2; pulses of O_2 had a similar effect. Sulphate did not promote proton extrusion; they attributed this to lack of ATP for sulphate activation in their experimental populations. In view of the apparent cytoplasmic location of APS reductase and sulphite reductase in *Desulfovibrio* (Badziong & Thauer, 1980; Odom & Peck, 1981*a*; Thauer & Badziong, 1981) it seems likely that the mechanism of chemiosmotic coupling which operates when sulphate is the electron acceptor is different from that with nitrate.

Odom & Peck (1981*b*, see also Peck & Le Gall, 1982) have proposed a 'hydrogen cycling' scheme for energy generation in *Desulfovibrio* which is compatible with the chemiosmotic scheme of Badziong & Thauer (1980): essentially, H_2 is released from organic substrates in the cytoplasm and permeates through the cell membrane to the periplasm. Here it is oxidized with the aid of cytochrome c_3, forming protons and generating electrons which are back-transferred across the cytoplasm for sulphate (or rather APS and sulphite) reduction. The proton gradient so formed would drive ATP synthesis in a conventional chemiosmotic manner. The elegance of this hypothesis is that, since the hydrogenase is periplasmic, H_2 produced by other organisms (or supplied by scientists) can equally generate a proton gradient, albeit in an unconventional direction.

An important experiment supporting the concept of hydrogen cycling is the observation by Odom & Peck (1981*b*; p. 63) that spheroplasts of *D. gigas*, which have lost all their periplasmic hydrogenase and cytochrome c_3, are unable to oxidize lactate. Ability to metabolize lactate was restored by adding purified hydrogenase + cytochrome c_3. The sensitivity of *Desulfovibrio* to acetylene inhibition, where *Desulfotomaculum* is insensitive (Payne & Grant, 1982), may be a related phenomenon since acetylene can be an inhibitor of uptake hydrogenase (Smith, Hill & Yates, 1976).

Metabolism of nitrogen

Ammonia is the principal source of nitrogen in conventional media for sulphate-reducing bacteria and this inorganic source is adequate. In the nomenclature of Lwoff (1944) these bacteria are allotrophs: they require organic carbon but not organic nitrogen. Senez, Pichinoty & Konavaltchikoff-Mazoyer (1956) regarded *Desulfovibrio desulfuricans* as freely permeable to ammonium ions, which is an unusual situation since most bacteria have an uptake mechanism for ammonia. Nitrite and hydroxylamine are utilized, as the same authors showed, probably by non-enzymic interaction with cytochrome c_3 (Senez & Pichinoty, 1958a, b).

Desulfobulbus propionicus reduces nitrate quantitatively to ammonia (Widdel, 1980; Widdel & Pfennig, 1982). Reports of nirate reduction by desulfovibrios appeared in the earlier literature but tended to be disregarded for two reasons. One was doubts about the purity of the cultures used, and the other was evidence that nitrate inhibits most cultures and enrichments containing desulfovibrios or desulfotomacula (van Delden, 1903; Iya & Sreenivasaya, 1945; Allen, 1949; Krasna & Rittenberg, 1955; Senez *et al.*, 1956). However, it now seems likely that dissimilatory reduction of nitrate to ammonia is not uncommon among desulfovibrios as long as certain conditions are met. Baumann & Denk (1950), using an isolate of desulfovibrio ('strain 3') which satisfied normal criteria of purity, described nitrate reduction accompanying sulphate reduction provided nitrate was below about 0.1% KNO_3; higher concentrations inhibited. Senez & Pichinoty (1958c) described a nitrate-reducing strain and McCready, Gould & Barendregt (1983) reported dissimilatory nitrate reduction by four desulfovibrio strains provided sulphate was absent. Liu & Peck (1981) studied a strain of *D. vulgaris* (American Type Culture Collection 27774) which grew anaerobically under argon with nitrate if its assimilatory sulphur requirements were satisfied by supplying cysteine and Na_2S. They isolated from it a new type of nitrite reductase which proves to be a 6-haem cytochrome of the *c* class of minimum mol. wt. about 66 000. It is autoxidizible. It reduces nitrite to ammonia and its high K_m for hydroxylamine suggests that this substance is not a free intermediate in the reaction. It may be able to reduce dithionite (Steenkamp & Peck, 1980); hydrogenase transfers electrons directly to the nitrite reductase without the intervention of cytochrome c_3 or an artificial carrier; Liu, DerVartanian & Peck (1980) detected an e.p.r. signal during turnover suggesting a haem–nitric oxide complex as an intermediate. As I mentioned above, energy-dependent proton translocation is associated with nitrite reduction and Steenkamp & Peck (1981) regarded the enzyme as located periplasmically.

Nitrogen fixation by sulphate-reducing bacteria was first claimed by Sisler & ZoBell (1951b) and substantiated by Le Gall, Senez & Pichinoty (1959). However, significant increments in fixed nitrogen or incorporations of $^{15}N_2$ are difficult to obtain in the laboratory and some doubt remained until Riederer-Henderson & Wilson (1970) confirmed fixation by several strains of *Desulfovibrio*, using the acetylene test as well as $^{15}N_2$. Postgate (1970a) confirmed nitrogen fixation by three out of five *Desulfovibrio* strains, using these tests, and also showed fixation by mesophilic desulfotomacula. Claims using the acetylene test that strains of *Desulfovibrio baculatus* and *D. africanus* fix N_2 (Nazina, Rozanova & Kalininskaya, 1979) regrettably lacked controls for endogenous ethylene production but can probably be accepted, but the significance of the much lower activities obtained with other strains is uncertain. Postgate's (1970a) thermophilic *Desulfotomaculum* strains did not fix N_2. Cell-free nitrogenase activity has been obtained in extracts of *Desulfovibrio desulfuricans* (Sekiguchi & Nosoh, 1973) but the activity was very weak. Dr T. H. Blackburn (quoted by Postgate, 1974) obtained a crude preparation of the molybdoprotein of nitrogenase from *D. desulfuricans* strain Berre S which would form a hybrid active nitrogenase with the iron-containing protein of the nitrogenase of *Klebsiella pneumoniae*.

As far as organic nitrogen metabolism is concerned, in *Desulfovibrio desulfuricans* Senez (1952) was unable to detect decarboxylation or deamination of aspartate, glutamate, α-alanine, arginine, glycine, histidine, isoleucine, leucine, lysine, ornithine, phenylalanine, proline, serine, tyrosine, valine, asparagine or glutamine. He did not test for oxidative decarboxylation with sulphate as hydrogen acceptor, though he examined glycine–leucine and arginine–glutamate systems for coupled oxido-reductions (Stickland reactions) and did not detect them. Acetone-dried powders, however, transaminated aspartic acid + α-ketoglutaric acid to glutamic acid and also α-alanine + α-ketoglutaric acid to glutamic acid. Senez & Cattaneo-Lacombe (1956) observed transaminase activity in a third strain as well as an ω-decarboxylase which converted L-asparagine to α-alanine.

Senez & Leroux-Gilleron (1954) examined the metabolism of sulphur-containing amino acids by a marine strain of *Desulfovibrio desulfuricans*. In the presence of sulphate, both cysteine and cystine were oxidatively degraded to acetate, CO_2, $NH_3 + H_2S$ but cystine was attacked only slowly and incompletely. In the absence of sulphate, cysteine was deaminated and the sulphydryl group removed, the products of degrada-

98 Metabolism

tion of the carbon skeleton were similar to those formed from pyruvate, suggesting the following sequence:

$$HS.CH_2CH(NH_2)COOH + H_2O \rightarrow CH_3CO.COOH + H_2S + NH_3$$

followed by:

$$CH_3CO.COOH + H_2O \rightarrow CH_3COOH + CO_2 + H_2$$

Cysteine biosynthesis does not follow the reverse course but takes place by way of O-acetyle serine (Gevertz, Amelunxen & Akagi, 1980).

An adenine nucleotide deaminase exists in a nitrogen–fixing strain of *D. desulfuricans* (Yates, 1969). An L-alanine dehydrogenase has been extracted from *D. desulfuricans* (Germano & Anderson, 1967, 1968) and studies on the biosynthesis of threonine and serine have been made (Daly & Anderson, 1966; Germano & Anderson, 1969).

Metabolism of phosphorus

The phosphorus metabolism of sulphate-reducing bacteria is somewhat unconventional. A special consequence of the need for 'energy-rich phosphate' in sulphate reduction seems to be the presence of a pyrophosphatase in *Desulfovibrio* which is unusual in requiring reduction to become active (Ware & Postgate, 1971). As explained earlier (p. 85), the enzyme 'pulls' the formation of APS from $ATP + SO_4^{2-}$. Since it becomes active only in reducing conditions, Ware & Postgate (1971) suggested that it conserved ATP by preventing futile formation of APS in aerobic conditions. Virtually no such enzyme is present in *Desulfotomaculum*; instead, as mentioned above (p. 94), this genus contains a pyrophosphate: acetate phosphotransferase (Liu & Peck, 1981) which conserves phosphate bond energy.

Phosphorylation reactions presumably accompany most anabolic and some catabolic reactions in sulphate-reducing bacteria and reactions such as the phosphoroclastic formation of acetyl phosphate from pyruvate are well established. Sorokin (1954*b*) reported that H_2 oxidation was accompanied by uptake of labelled phosphate, consistent with ATP generation, and that it was released when CO_2 was introduced. Though based on the now discarded view that these bacteria are autotrophic, his finding is still valid in its suggestion that anabolic reactions can be accompanied by hydrolysis of stored ATP.

Metabolism of iron

Desulfovibrio shows an exceptionally high requirement for inorganic iron (Postgate, 1956*b*). The iron is needed for such cell constituents as ferre-

Metabolic inhibitors

doxin and cytochromes c and, in iron-starved cultures, the content of cytochromes c is less. Tsuji & Yagi (1980) briefly mentioned that it is cytochromes other than c_3 which are depleted in conditions of iron limitation: Postgate (1956b) had assumed that the decreased absorption at 553 nm by iron-starved cells was attributable to lack of cytochrome c_3 because the other cytochromes c had not then been discovered. *Desulfovibrio desulfuricans* grown on pyruvate without sulphate contains less of the cytochromes c and has a lower absolute Fe^{2+} requirement. Desulfoviridin contents show only slight changes in Fe-starved conditions but the hydrogenase content of iron-starved cells is low. The plentiful supplies in *Desulfovibrio* of iron-rich proteins such as ferredoxins, cytochromes of the c_3 group, even nitrogenase in some strains, must present physiological problems as far as the uptake of Fe^{2+} is concerned, because S^{2-} is a normal metabolic product and the solubility of FeS is miniscule. It is likely that the siderochromes (Nielands, 1973) of these bacteria, if they or their equivalents exist, have a remarkable affinity for iron; iron uptake in these bacteria would repay study.

Metabolic regulation

There has been considerable scientific interest in both genetic regulation of enzyme synthesis and physiological regulation of enzyme function in many types of micro-organism, but the special difficulties of handling sulphate-reducing bacteria have so far delayed extension of such studies to these organisms. So little is known of their genetics that genetic regulation has no experimental basis; studies of physiological regulation are also few but some interesting features may be expected in this area because of the obligately anaerobic nature of these bacteria, yet their ability to survive oxygen stress. There is evidence that, in *Desulfovibrio*, exposure to oxygen leads to modification of the soluble hydrogenase (p. 72) and the pyrophosphatase (p. 98) *in vivo*, both of which changes may have regulatory functions by preventing futile metabolism in inappropriate (i.e. aerobic) conditions. Phosphorylation or dephosphorylation of the flavodoxin of *D. gigas* may determine whether vectorial electron transport or substrate level phosphorylation is its major source of ATP (p. 81).

Metabolic inhibitors

Growth inhibitors were discussed in Chapter 2 (p. 47) and are collected in Appendix 2. Specific antagonists of sulphate reduction were discussed

100 *Metabolic inhibitors*

earlier in this chapter (p. 83). Reports of use of conventional metabolic inhibitors with sulphate-reducing bacteria have been rather few, but Sorokin (1954*b*) observed that sodium azide at 0.1 to 1 µmol/ml inhibited growth of *Desulfovibrio* while actually stimulating the rate of sulphate reduction in H_2. This observation led him to suspect that phosphate uptake might be coupled with sulphate reduction, and he demonstrated that azide acted by inhibiting the uptake of phosphate by cell suspensions reducing sulphate. Cyanide at 1 to 5 µmol/ml had a similar effect. Ishimoto *et al.* (1954*b*) studied the effect of several inhibitors, noting that, in hydrogen, sulphate reduction by cell suspensions was strongly inhibited by cyanide, selenate or arsenite, but that phenosafranine reduction in H_2 or methylene blue reduction with lactate were little affected. Sulphite and thiosulphate reduction were intermediate in sensitivity. Azide, hydroxylamine and tungstate also had inhibitory effects. Viologen dyes have a cytochrome-like action with preparations from *Desulfovibrio* and this may explain their pronounced effect on the metabolism of undamaged cells (*Chemistry Research*, 1956; Peck, 1960). Methyl and benzyl viologens strongly inhibit sulphate reduction by resting cells; thiosulphate and sulphite reduction are not so influenced. The effect of chromate or molybdate in depleting the ATP reserves (Taylor & Oremland, 1979) was mentioned earlier in this chapter (p. 84). Acetylene, through its action on hydrogenase, is a metabolic inhibitor for *Desulfovibrio* (Payne & Grant, 1982; see also p. 95).

6
Evolution

The evolutionary relationships within the sulphate-reducing bacteria present a fascinating problem. There is no reason to regard the genera *Desulfovibrio, Desulfotomaculum, Desulfobulbus, Desulfobacter* etc. as phylogenetically related. Even within *Desulfovibrio*, which is perhaps the best analysed genus, knowledge of the properties of the species is insufficiently detailed for them to be arranged in evolutionary hierarchies. The existence of three clusters of DNA composition in the genus *Desulfovibrio* (see Chapter 2) might indicate considerable evolutionary divergence within the genus, consistent with its being long established. On the other hand, the apparent diversity of prokaryote types which are now known to be capable of sulphate reduction suggests that, as with nitrogen fixation (see Postgate, 1982*b*), the genetic information for sulphate reduction forms a linkage group which has undergone transfer between diverse anaerobic genera fairly readily during evolutionary history. Genetic studies with these organisms are now badly needed, for the discovery of clustered or plasmid-borne sulphate reduction genes would greatly influence thinking on the evolution of this group. The writer, with Dr R. Robson, has observed that two cryptic plasmids (about 40 000 and 80 000 mol. wt) occur naturally in *Desulfovibrio gigas* but could find none in strains of *D. desulfuricans, D. salexigens* and *D. africanus*.

Little can be said about the phylogenetic relationships of sulphate-reducing bacteria to other bacterial groups. *Desulfotomaculum nigrificans* was once classified with the genus *Clostridium* but even then its Gram-negative reaction and thermophily made it unusual; *Desulfovibrio* has been classified with *Vibrio* and *Spirillum*, but the sole basis was their curved morphology and phylogenetically the genera have nothing in common. *Desulfonema* has superficial resemblances to *Beggiatoa* but the presence of desulfoviridin in *D. limicola* might in fact bring it closest to *Desulfovibrio*. A similar argument might apply to *Desulfococcus*, which also gives a positive desulfoviridin test. However, as I indicated in Chapter 2, too few definitive characters are yet available for a rational phylogeny to be constructed.

Dissimilatory sulphate reduction has not been recorded among the Archebacteria, although dissimilatory sulphur reduction occurs within this group (p. 3).

The peculiar metabolism of the sulphate-reducing bacteria has had an important bearing on current thinking about the early development of living things on this planet. The most primitive denizens of the planet were probably heterotrophic anaerobes resembling present-day *Clostridium* and *Bacteroides*: organisms subsisting on fermentation of photochemically generated organic matter (Bernal, 1967; Rutten, 1972). As the 'primitive soup' became exhausted, natural selection would favour the emergence of chemotrophic and autotrophic anaerobes (see also Broda, 1975). There are reasons to believe that free sulphate was available on a substantial scale before much free O_2 was present. One reason is that sulphide was common in the primitive environment and would have 'mopped up' O_2, becoming converted to sulphate itself, by ordinary chemical reaction. Another is that anaerobic photo-oxidation of sulphide to sulphate by prototypes of the modern coloured sulphur bacteria might have contributed sulphate to the environment. Sulphite from volcanic SO_2 would also have been present. CO_2 was available, so the most probable chemotrophs for the next phase of terrestrial evolution were sulphate-reducing, carbonate-reducing and photosynthetic sulphide-oxidizing anaerobes. As regards sulphate reduction, there exists some evidence that this view is correct: the relative abundance of the natural isotopes of sulphur becomes altered during sulphate reduction (see p. 89), the sulphide being enriched in the lighter isotope ^{32}S compared with the heavier ^{34}S which remains as sulphate. This fractionation has been abundantly confirmed (e.g. Jensen, 1962) and serves to distinguish sulphide- or sulphur-containing minerals which have been formed by biological action from those which have not. Ault & Kulp (1959) showed that sulphur-isotope fractionation indicative of biological sulphate reduction can be detected in rocks as old as 2×10^9 years, well into the Precambrian era; Thode (1980) and his colleagues (see also Monster *et al.*, 1979) detected evidence of biological fractionation 2.7×10^9 years ago, but not at 3 to 3.7×10^9 years ago, which sets limits to the date of appearance of massive biological sulphate reduction on this planet. Detection of isotope fractionation in ancient deposits does not necessarily imply that sulphate was the substrate; in a discussion of these matters, Skyring & Donnelly (1982) suggested that sulphite reduction was the most primitive process, responsible for a very early 'burst' of isotope fractionation, and that ecologically significant sulphate reduction developed later.

Sulphur-isotope fractionation attributable to biological processes, probably including sulphate reduction, was thus an important process in the biosphere well before the atmosphere contained significant amounts of free O_2 (1 to 1.6×10^9 years ago according to Rutten, 1972; no earlier than 2×10^9 years ago according to Schidlowski, 1980). At this period there were probably no eukaryotic organisms (conventional fossils are not found earlier than 0.5×10^9 years ago) but simulacrae of micro-organisms resembling cyanobacteria have been detected in rocks of comparable age.

So we have direct evidence of biochemical processes akin to present-day sulphate reduction being important in the earliest stages of life on this planet and since, as will be discussed in Chapter 8, most of the world's sulphur deposits have been formed as a result of massive sulphate reduction, there have clearly been periods when sulphate reduction was an important, and possibly dominant, vital process.

As with many present-day representatives of primitive creatures, it is tempting to look for primitive characters remaining in today's sulphate-reducing bacteria. These questions have been discussed by Peck (1966–7), Klein & Cronquist (1967), Postgate (1968), Peck (1974) and briefly summarized by Le Gall & Postgate (1973). Sulphate-reducing bacteria have several properties which appear to be primitive, even compared with other bacteria, and these include:

1. Possession of hydrogenase linked to a phosphoroclastic breakdown of pyruvate.

2. Use of ferredoxin and rubredoxin as electron-transport factors. Rao & Cammack (1981) considered that the structures of *Desulfovibrio* ferredoxins related them closely to an archetypal clostridial ferredoxin, from which both they and the ferredoxins of coloured sulphur bacteria evolved. Schwartz & Dayhoff (1978), judged from the sequence of a ferredoxin from *Desulfovibrio gigas* compared with sequences of other bacterial ferredoxins, that *Desulfovibrio* diverged from the prokaryotic evolutionary tree later than phototrophic sulphide-oxidizing bacteria. On the other hand, Vogel, Bruschi & Le Gall (1977) assigned *D. gigas* and *D. vulgaris* closer to *C. pasteurianum* on the basis of the comparative sequences of their rubredoxins.

3. Ability to conduct a reductive carboxylation of acetate to form pyruvate. This process is considered to be a primitive CO_2 assimilation process (see Kerscher & Oesterhelt (1982).

4. The possession of the mixotrophic assimilation processes discussed

in Chapter 3, which seem to represent an evolutionary link between heterotrophy and true autotrophy (the ability to assimilate CO_2 alone). Sorokin's reaction – the mixotrophic assimilation by *Desulfovibrio* of CO_2 together with acetate under H_2 (see Chapter 3) – recalls the ability of certain primitive photosynthetic bacteria to photo-assimilate acetate in place of CO_2.

5. The possession of cytochromes as well as ferredoxins might suggest an intermediate evolutionary status between the aerobes and the anaerobes.

6. The pathway of synthesis of citrate is, in certain strains, of a character associated with primitive types of bacteria (Gottschalk & Barker, 1967) and so is a side reaction of the pyruvic phosphoroclasm which leads to the formation of small amounts of methane (Postgate, 1969b; see also Chapter 5). These reactions may be 'biochemical vestiges': reactions which were relevant in the early evolutionary history of the organism but which are now trivial.

7. Han & Calvin (1969) have shown that the hydrocarbons of *Desulfovibrio* are typical of extremely primitive microbes.

8. The sirohydrochlorin of desulfoviridin may be 'primitive' as compared with the other porphyrin molecules (see Siegel, 1978; Skyring & Donnelly, 1982).

9. Dickerson, Timkovich & Almassy (1976) observed that the folding pattern deduced from X-ray data on cytochrome c_{553} of *Desulfovibrio* (not cytochrome c_3) related it to several other mono-haem bacterial cytochromes and the authors placed it very early in the hierarchy of evolution of c-type cytochromes.

10. The possession of an iron-containing superoxide dismutase (Hewitt & Morris, 1975; Morris, 1975) may be a primitive feature since this type of enzyme may have preceded the copper and manganese enzymes of most bacteria. However, superoxide dismutases are not uniformly found in *Desulfovibrio*, so the enzyme may instead have been a late acquisition; the question was discussed by Hatchikian, Le Gall & Bell (1977).

11. Comparison of sequences in the ribosomal RNAs of *Desulfovibrio* species suggests that they branched from an hypothetical 'ancestral eubacterium' at an early stage in the evolution of prokaryotes, although later than the Archebacteria (Fox *et al.*, 1980; Stackebrandt & Woese, 1981). These studies tend to place *Desulfovibrio* later than green sulphide-oxidizing bacteria but in a 'cousin' relationship to red sulphide-oxidizing bacteria, both of which represent very early prokaryotic types.

Thus one can point to a number of properties of present-day sulphate-reducing bacteria which seem primitive, and a decade ago one would have been tempted to set up an evolutionary hierarchy among prokaryotes with sulphate-reducing bacteria near the origin. However, present understanding of gene transfer among microbes, in particular the recognition that large 'packages' of genetic information can become transferred among prokaryotes (as transposable elements, on plasmids or by other transformation processes), leaves considerable doubt whether present-day prokaryotes bear any coherent evolutionary relation to their early ancestors at all. Today, it is probably wisest in considering microbial evolution to disregard existing species and to consider processes, enzymes or even genes (although in the present context we have no relevant information on the latter). Yet the consensus of views on the evolutionary status of sulphate-reducing bacteria is still reasonable: that they represent early evolutionary steps towards a cytochrome-based aerobic metabolism, away from obligate fermentative anaerobes. They also represent a simultaneous departure from purely heterotrophic metabolism, yet they mostly did not solve the problems of a wholly autotrophic existence; they probably emerged after the most primitive anaerobes and the Archebacteria, which includes the methanogenic bacteria. It now seems rather unlikely, if *Desulfovibrio* is a valid guide, that sulphate-reducing bacteria could have been precursors of the phototrophic sulphur bacteria (see Klein & Cronquist, 1967; Peck, 1966–7). They may have been 'cousins' on a different evolutionary branch or, as Peck (1974) suggested and Pfennig & Widdel (1982) accepted, they may have been preceded by sulphide-oxidizing phototrophs.

In considering these questions, one must bear in mind that there is little reason to believe that the early types of either group bore much, if any, phylogenetic relationship to their present-day counterparts. For example, the present-day phototrophs show some relatively sophisticated metabolic patterns which probably have little bearing on their antiquity as a group. This point, of course, applies to sulphate-reducing bacteria – and to many other microbes. In later eras associations of sulphate-reducing bacteria with phototrophic and autotrophic sulphide-oxidizing bacteria may well have provided an early primary producing process for CO_2 fixation before conventional O_2-evolving photosynthesis became established (see Postgate, 1968).

In the context of the ancient character of sulphate-reducing bacteria a paradox arises which was briefly mentioned by Ware & Postgate (1971). Many of their electron-transport components react spontaneously with oxygen and, indeed, desulfovibrios are capable of a degree of aerobic metabolism. Yet they are very exacting anaerobes as far as growth is concerned. Present knowledge offers no reason why, for example, such organisms could not grow aerobically at the expense of, say, pyruvate, using only substrate level ATP for growth; the presence of catalase and superoxide dismutase (Hatchikian, Le Gall & Bell, 1977) excludes explanation in terms of inadvertent peroxide formation. Yet neither aerobic nor facultative sulphate-reducing bacteria are known – yet.

7
Ecology and distribution

The ecology of an organism is the science of its interaction with its environment. It includes the study of (*a*) the environmental conditions which permit or favour growth and multiplication of the organism, (*b*) the effect of growth and multiplication of the organism on that environment, and (*c*) the interactions between the organism and other living things which collectively constitute the eco-system. As applied to micro-organisms, ecology and distribution do not necessarily correspond: owing to their capacity to remain dormant, microbes can frequently be isolated from environments which seem totally unsuited to them and on which they seem to be having no effect. The sulphate-reducing bacteria provide a particularly good example of this paradox because of their almost universal distribution but limited ecology (a converse situation occurs with the photosynthetic bacteria which often do not occur when one would expect them to). Systematic study of their ecology and distribution was delayed by two main difficulties. The first is a practical one: reliable methods for the determination of numbers of sulphate-reducing bacteria are few and they have only been established as reliable for a few species of *Desulfovibrio*. Hence it has often been difficult to support any proposal that they were active in a given environment by a demonstration of an abnormally high population there. Despite their economic importance, until recently no-one had any clear idea what a *normal* population of sulphate-reducing bacteria might be, though, if the bacteria were not easily demonstrated in a fresh field sample it was reasonable to conclude that they were not very active there. The second difficulty might be termed a psychological one: in those environments in which sulphate-reducing bacteria were indisputably active, their efforts were usually economically so important that attention was naturally diverted from the mechanism of their effect on the environment to the exploitation or prevention of that effect.

For reasons such as these the ecology of these bacteria was rather neglected for many years, with little more than descriptive data being

108 Ecology and distribution

published. However, when radio-isotopes became available for ecological studies it became practicable to use ^{35}S, a low energy β-emitter, as a means of studying sulphate reduction (along with other steps in the sulphur cycle), in an eco-system, without actual knowledge of the bacteriology of the processes. In sediment cores or water samples, for example, Na$_2^{35}$SO$_4$ could be introduced with minimal disturbance and the kinetics of its conversion to sulphide or sulphur could be followed quantitatively (see Ivanov, 1968). The methodology of such analyses was examined by Jørgensen (1978a, b, c); tracer work has underpinned most of the more recent discoveries on the ecology of sulphate reduction, particularly when coupled with studies of ^{14}C turnover.

Distribution

Traditional methods had early demonstrated that, as a group, the sulphate-reducing bacteria include terrestrial and aquatic types, marine and strongly halophilic types, thermophilic and psychrophilic types, sporulating and non-sporulating types (ZoBell, 1958). Moreover, though the complete interconversion of thermophilic and mesophilic species mentioned in Chapter 1 does not take place, the organisms show considerable adaptability in terms of temperature (Campbell *et al.*, 1956; Postgate, 1956c; Campbell & Postgate, 1965) and salinity (van Delden, 1903; Baars, 1930; Littlewood & Postgate, 1957a; Ochynski & Postgate, 1963). Representatives of the sulphate-reducing bacteria are to be found in almost all ordinary environments on this planet: they are present in soils (Young, 1936); fresh, marine and brackish waters; artesian waters; hot springs and geothermal areas (Kaplan, 1956); oil and natural gas wells; sulphur deposits; estuarine muds; sewage; salt pans (Saslawsky & Chait, 1929); corroding iron; the rumina of sheep (Lewis, 1954) and guts of insects (see Postgate, 1960a). They tolerate temperatures from below -5 to 75 °C and show considerable adaptability to new conditions of temperature; they can be grown *in vacuo* or in water under a pressure of 1×10^5 kPa; they tolerate pH values ranging from below 5 to 9.5 and a wide range of osmotic conditions; in the Solar Lake of Sinai they tolerate salinities up to 18% and temperatures from 16° to 60 °C (Jørgensen & Cohen, 1977), but no truly osmophilic sulphate-reducing bacteria have been obtained in laboratory culture in hypersaline media. A remarkable environment, a supercooled, brackish pool in the Antarctic, was described by Barghoorn & Nichols (1961); ZoBell (1958) pointed out that probably more sulphate-reducing bacteria in nature function at below 5 °C than above, because of their abundance on the ocean beds. Organisms from deep waters may be barophilic: they grow best at high pressures and some need such pressures.

The terrestrial record for growth in extreme conditions was for long held by a strain of sulphate-reducing bacteria obtained from a deep drilling in an oil or sulphur deposit: it grew above the normal boiling point of water under high pressure (104 °C at 1×10^5 kPa hydrostatic pressure, ZoBell, 1958). However, Baross & Deming (1983) have swept the board with their demonstration of barophilic thermophiles in deep Pacific hydrothermal vents capable of multiplying, under pressure, at over 250 °C. *Desulfovibrio* is the most common inhabitant of marine, brackish and estuarine environments and authentic, truly thermophilic strains (able to survive repeated subculture at above 50 °C) of this genus, believed to be non-existent for two decades, are now known (Rozanova & Khudyakova, 1974). It is curious that *Desulfotomaculum* is not normally present in marine or saline environments (cf. p. 9), because laboratory strains acclimatize readily to saline conditions. The thermophilic sulphate-reducing bacteria are usually strains of the species *Desulfotomaculum nigrificans*; they are present in, and probably responsible for sulphide formation in, deep telluric aquifers subject to geothermal heating (Olson *et al.*, 1981). Soils, the rumina of ruminant mammals and the guts of insects can contain mesophilic strains of either *Desulfovibrio* or *Desulfotomaculum*; *Desulfotomaculum acetoxidans* is often found in faeces or the intestinal tract (Widdel & Pfennig, 1977) but other sulphate-reducing bacteria may also be present. In soils desulfotomacula are often dominant (e.g. Adams & Postgate, 1961). Newly polluted environments, or environments which are regularly polluted, seem to provide conditions which favour *Desulfovibrio*.

Sulphate-reducing bacteria are thus adaptable to almost any natural environment on this planet except, strangely enough, the most common: an ordinary aerobic environment. Their need for a low redox potential for multiplication restricts their activities to reducing environments. However, they can be isolated from almost any soil or water sample taken at random so they can survive long exposure to oxygen and become active again if and when they return. The ability of *Desulfovibrio vulgaris* to survive exposure to O_2 was documented by Hardy & Hamilton (1981). Sulphate-reducing bacteria are rarely encountered as airborne organisms, though both sporulating and non-sporulating strains survive drying in soils.

Environmental effects of sulphate reduction

Once growth of sulphate-reducing bacteria starts in a given environment, the chemical and physical nature of that environment changes, sometimes drastically. It is useful to analyse the various environmental effects that massive biological sulphate reduction may have on an eco-system.

Formation of sulphide: the end-product of sulphate reduction is the sulphide ion, which has an equilibrium E_m at 15 °C and pH 7 in the region of -320 mV. In such an environment free O_2 can only exist transiently except at very low concentrations,[1] consequently anaerobic organisms will become active and the activities of many aerobes will cease. The aerobic microflora of the environment will tend to be replaced by anaerobes and organic material will be fermented rather than oxidized.

Besides suppressing the multiplication of aerobic micro-organisms, sulphide is toxic to many of them, as well as to all but a few macro-organisms. Consequently most higher animals and plants will be killed and putrefaction of their corpses will augment the organic material available to the sulphate-reducing bacteria.

Alteration of pH: the alkali and alkaline earth sulphides dissociate in solution to yield free H_2S as well as HS^- and OH^- ions. Since H_2S is volatile the pH of the environment thus tends to become alkaline. Over a long period this alkalinity is neutralized by atmospheric CO_2, so that the carbonate and bicarbonate accumulate; during periods of active sulphate reduction, however, the environment tends to become alkaline unless compensating metabolic reactions leading to acid formation are taking place simultaneously, or unless the sulphide is trapped as insoluble derivatives of heavy metals.

Removal of heavy metals: heavy metal ions become precipitated as the metal sulphide. Since soluble iron is fairly common in soils and waters, blackening due to FeS formation is a characteristic of activity of sulphate-reducing bacteria. In theory, precipitation of iron could lead to biological iron deficiency in waters, and at least one such case has been proposed (p. 113). Since phosphate can be bound in soil as ferric phosphates, con-

---

[1] That it can nevertheless exist in sulphide solutions was demonstrated ingeniously by Beijerinck (1904) who showed that H_2S water could provoke O_2-dependent luminosity when added to the bottom of a culture of luminous bacteria.

version of this to FeS can sometimes augment fertility by making phosphates available to plants.

Removal of organic matter: sulphate reduction requires a substantial consumption of organic matter which is converted, sometimes via acetate, to CO_2. Thus a net mineralization of organic matter takes place anaerobically. Sulphate-reducing bacteria can also contribute to mineralization in low sulphate environments, when methane $+CO_2$ are the products, as in a tripartite cellulose-decomposing community described by Laube & Martin (1981).

Removal of hydrogen: H_2 is a characteristic product of anaerobic fermentations, and the presence of hydrogenase in the majority of the acetate-producing sulphate-reducing bacteria enables them to utilize H_2 for sulphate reduction. Utilization of cathodic H_2 probably contributes to the anaerobic corrosion of iron by these bacteria. Scavenging of H_2 has been suggested as one interpretation of the apparent incompatibility of sulphate-reducing and methane-producing bacteria (p. 121).

Removal of sulphate: sulphate reduction in polluted river or pool water usually ceases because the sulphate is exhausted rather than because all the organic material has been used. In the sea, sulphate is plentiful but marine sediments can become depleted of sulphate.

Alteration of commensal microflora: the balance of even anaerobic flora is altered by massive growth of sulphate reducers. For example, methane formation and sulphate reduction are incompatible on a significant scale (see below, p. 120). Brackish waters sometimes contain coloured non-sulphur bacteria and green algae; if massive sulphate reduction takes place these tend to be replaced by coloured sulphur bacteria and cyanobacteria. In certain circumstances growth of aerobic thiobacilli may be favoured. These changes can culminate in the establishment of the 'sulfuretum' to be discussed shortly (p. 116).

Nitrogen fixation: in most terrestrial environments the contribution of *Desulfovibrio* to the nitrogen status of the environment is probably trivial. However, *Desulfovibrio* species seem to be the only well-established nitrogen-fixing heterotrophic inhabitants of truly marine sediments and are the dominant nitrogen fixers in lakes with marine substrata.

112   Ecology and distribution

Isotope fractionation: though environmentally trivial, the fractionation of sulphur isotopes which accompanies sulphate reduction should be mentioned here for completeness. Its use in an evolutionary context was discussed in Chapter 6.

Establishment of a geochemical oxidant sink: as well as having the biological consequences mentioned above, the reducing nature of the microenvironments generated as a result of sulphate reduction leads to cumulative removal of oxidants. In particular, Board (1976) suggested that sulfureta (p. 116) play a part in regulating the O_2 content of the earth's atmosphere.

Populations in natural environments

Reliable quantitative data on numbers of sulphate-reducing bacteria in natural environments are hard to come by, because many of the numbers quoted in the scientific literature were obtained using culture media of unsatisfactory E_h (see Chapter 3). Kimata, Kadota, Hata & Tajima (1955*a*, *b*, *c*) studied the distribution of sulphate-reducing bacteria in Hiroshima Bay, determining the bacteria quantitatively using media poised with ascorbate or Na_2S. Their data may be taken as reliable as far as *Desulfovibrio* is concerned since it gave theoretical recoveries (about 10^9 viable cells/ml) with a pure culture (Kimata *et al.*, 1955*a*, their Table 4 on their page 232). The sea water contained 1 to 10 viable desulfovibrios/ml irrespective of the depth or site of the sample, but the bottom muds contained considerably greater populations: 10^2 to 10^5 viable cells/g. The population in the mud was broadly proportional to (*a*) the sulphide content, and (*b*) the biological oxygen demand. Thus there appeared to be a direct dependence of activity on the organic content of the mud, a dependence brought out by the fact that muds near to the outlet of a paper mill often had high contents of organic material and sulphate-reducing bacteria. In a later paper they showed that the desulfovibrios needed the organic material as a source of carbon rather than nitrogen (Kimata *et al.*, 1955*a*) Workers from the laboratory also examined a polluted estuarine environment (Kimata *et al.*, 1958) and came to the conclusion (*a*) that the population of the water consisted of organisms disturbed from their natural habitat, the bottom mud, and (*b*) that the total numbers of desulfovibrios in the bottom mud were little influenced by the organic content, sulphate content or salinity of the mud, though the ratio of fresh-water to salt-water types depended on the salinity of the mud. The sulphide content of the mud depended on both its sulphate content and its content of organic matter. Kadota & Miyoshi (1964) provided further details of their ecological studies. Koyama & Sugawara (1953) reported that the bottom muds of a brackish lake showed limited sulphate reduction where it would normally be expected and attributed this phenomenon to iron deficiency enhanced by the presence of Mn^{2+} which acted antagonistically to the limited amount of iron present. Hines & Buck (1982) found a close correlation between sulphate concentration and population of desulfovibrios, using poised media, in the upper 10 cm or less or near-shore marine sediments (10^4 to 4×10^4/g); neither desulfovibrios nor sulphate persisted below this depth. Methanogens were distributed fairly evenly to a much greater depth (~ 50 cm).

Desrochers & Fredette (1960) used a reliable but now obsolete procedure to demonstrate a dramatic increase in the count of *Desulfovibrio*, from 10^2/ml to about 10^6/ml, in the Ottawa river (Canada) after the point at which a paper mill effluent was admitted to the river. Lighthart (1963) showed a modest increase (10^4 to 10^5/g) of sulphate-reducing bacteria in the sediment of San Vicente Reservoir, California, which accompanied a disappearance of free O_2 from the ambient water and a drop in E_h from about -30 to -150 mV. Cappenberg (1974b) used poised media and found virtually no desulfovibrios in the water of Lake Vechten (Holland) except during summer stratification when a small number (~ 10/ml) were detectable at depth. Larger numbers (~ 200/ml) resided at the surface of the bottom mud and seasonal fluctuations occurred.

Sulphate-reducing bacteria are often to be found in the intestinal tract. *Desulfotomaculum ruminis* and *D. acetoxidans* are both of intestinal origin. Howard & Hungate (1976) recorded that the population of *Desulfovibrio* in the sheep rumen is about 10^8/ml, a number consistent with the rate of sulphide formation in that environment.

Soil microbiologists distinguish two classes of microflora in an environment: the autochthonous microbes which are usually present and which are often plentiful, and the zymogenous population which are microbes whose numbers fluctuate rapidly in response to environmental change. Though *Desulfovibrio* is by far the easiest organism to obtain from natural environments, experience has for many years suggested that it may be more of a zymogenous than an autochthonous type: in a crude survey, Adams & Postgate (1961) found an unexpected prevalence of mesophilic desulfotomacula (then recorded as *Desulfovibrio orientis*) in soil samples. The situation was elegantly quantified by Jørgensen (1978c) who compared rates of reduction of labelled sulphate in a temperate coastal marine sediment with the population of desulfovibrios as estimated by colony counts, using a poised medium. He concluded that, though there was some proportionality between apparent population and metabolic activity, the counting procedure must have underestimated the true population by up to 1000-fold. Sulphide formation underestimated the scale of sulphate reduction by about 10-fold, because of diffusional losses in the sediment. The conclusion is compelling that *Desulfovibrio* is a minor component of the normal population of an undisturbed sediment. Consistent with this view, Laanbroek & Pfennig (1981) found *Desulfobacter*, *Desulfobulbus* and *Desulfovibrio* in the approximate ratio of (6 : 51 : 22) $\times$ 10^4/ml of marine sediment. *Desulfobacter* was absent from a fresh-water sediment and *Desulfobulbus* : *Desulfovibrio* were (24 : 16) $\times 10^4$/ml. It is possible that their numbers for *Desulfobacter* and *Desulfobulbus* were underestimates, because they reported no check that their counting procedure recovered viable organisms of these genera quantitatively.

Such studies indicate strongly that the autochthonous sulphate-reducing population of many natural environments is probably not *Desulfovibrio*. Since present bacteriological techniques count only this genus reliably, numerical data on populations of the autochthonous types are not readily available and, conversely, counts of *Desulfovibrio* are of very limited use in this context. This is not to decry the value of such counts in situations associated with zymogenous multiplication of sulphate-reducing bacteria: many of the situations to be discussed in Chapter 7 result from massive multiplication of sulphate-reducing bacteria in response to pollutants or other factors. In such circumstances, desulfovibrios are likely to dominate and counts of them can then be very valuable.

For these reasons, much of our present understanding of sulphate reduction has been obtained from studies of sulphate reduction as a process, using chemical or isotopic measurements, without a precise knowledge of the microbiology of the system. These matters will shortly be discussed, but another aspect of their microbiology will be indicated first.

Consortia and syntrophs

Sulphate-reducing bacteria form relatively stable associations with certain other sulphur bacteria, based on nutritional inter-dependence. There is no doubt at all that syntrophic associations of sulphur bacteria occur frequently, and in many varieties, in nature, but whether more determinate consortia actually exist is still not established. One of the best-studied examples is the case of the so-called 'Chloropseudomonas ethylica', which was believed to be a new type of sulphide-oxidizing phototroph when first isolated and was later thought to be a consortium of Chlorobium with Desulfovibrio (see Gray, et al., 1973; Gray, 1977); it now seems likely that the heterotrophic partner of the consortium was Desulfuromonas acetoxidans and the phototrophic partner was sometimes Chlorobium limicola, sometimes Prosthecochloris aestuarii, both sulphide-oxidizing phototrophs. Other consortia of phototrophic sulphur bacteria with heterotrophs such as Chlorochromatium or Pelochromatium (see Starr & Schmidt, 1981) may yet prove to involve sulphate-reducing bacteria as the sulphide-generating partner (Pfennig, 1978).

An important class of association in which sulphate-reducing bacteria participate involves the transfer of H_2 between partners. McInerney & Bryant (1981) showed that mixed cultures of a hydrogen-consuming methanogenic bacterium (Methanobrevibacter smithii, earlier Methanobacterium ruminantium) with Desulfovibrio desulfuricans or D. vulgaris fermented lactate to acetate + methane provided that the sulphate concentration was low. They attributed the reaction to the sulphate-free dismutation of lactate to yield free H_2 (p. 63), a reaction which is thermodynamically unfavourable and which ceases forthwith unless another organism, in this case the methanogen, scavenges the H_2 at once. A co-culture of Desulfovibrio with Methanosarcina barkeri, which can utilize both H_2 and acetate, converted lactate quantitatively to methane $+CO_2$; such co-cultures can take up to two months to complete growth. Sulphate interferes with such syntrophic associations (Bryant et al., 1977) by preventing significant H_2 evolution from lactate.

In the type of syntrophy just discussed, the Desulfovibrio partner generates H_2 for the methanogen. Another class of syntrophic system exists in which Desulfovibrio 'pulls' a thermodynamically unfavourable reaction by scavenging H_2; the K_m of Desulfovibrio hydrogenase is very low so it is an effective partner in such a system. Recognition of this type of system has led to the discovery of obligatory syntrophic bacteria. These are obligate anaerobes which can oxidize highly reduced carbon sources, such as long chain fatty acids, by removing H_2 from the molecule, but only if the reaction is 'pulled' by a desulfovibrio or a methanogen. Several examples have been described. Syntrophobacter wolinii ferments propionate to yield sulphide + acetate in co-culture with desulfovibrio, or methane + acetate in co-colture with a methanogen (Boone & Bryant, 1980); Syntrophomonas wolfii degrades butyrate to give similar products according to the nature of its syntrophic partner (McInerney et al., 1981; McInerney,

Mackie & Bryant, 1981); an organism described by McInerney, Bryant & Pfennig (1979) ferments butyrate, caproate or caprylate in co-culture with a hydrogenase-containing sulphate-reducing or methanogenic partner. In these systems the primary degradation of the fatty acid is formally endergonic ($\Delta G = +8$ to $+40$ kJ/mole) but it becomes exergonic provided the hydrogenase activity of the partner holds the P_{H_2} down to about 10^{-6} atm ($\Delta G = -4$ to -8 kJ/mole). The overall reaction may then have a larger energy yield: that for *Syntrophobacter wolinii* + *D. desulfuricans* is about $\Delta G = -38$ kJ/mole propionate consumed. Boone & Bryant (1980) tabulated thermodynamic data for syntrophisms of this kind. Such syntrophic systems may well be extremely important in microbial ecology, and their discovery has led to renewed interest in the general role of syntrophs and consortia in microbiology. It is necessary however to mention some reservations which persist: firstly, it ought to be possible to culture the 'obligate' syntrophs in pure culture with a non-living H_2-scavenging system (e.g. hydrogenase or colloidal palladium plus an electron acceptor); secondly, actual hydrogen transfer between the species has yet to be demonstraed with ^2H or ^3H isotopes as tracers. Finally, it is curious that two-partner syntrophy should survive in the natural environment when one would expect more effective energy conservation if both reactions were to take place within one organism. As an example one can cite *Desulfobulbus propionicus* which conducts essentially the same reaction as does the partnership of *Syntrophobacter wolinii* + *Desulfovibrio*. Perhaps the ability to function with more than one partner has survival value for the obligate syntroph.

Role in the sulfuretum

Massive sulphate reduction in nature always results from the activities of bacteria and is often the penultimate stage of gross organic pollution of an environment. A typical example would be a temperate woodland pond in the autumn. In summer, such a pond would be reasonably well aerated and its microbial population might be limited by inorganic nutrients, supporting modest populations of algae, commensal bacteria and actinomycetes. Leaf fall in autumn would provide an abrupt influx of organic matter, causing rapid growth of aerobic bacteria. Oxygen would then become depleted, aerobic organisms would die and augment the pollution, and so facultatively anaerobic denitrifying bacteria would flourish briefly until nitrate became exhausted. Then the sulphate-reducing bacteria would multiply, augmenting pollution further through the toxic action of the H_2S they form. Re-oxidation of H_2S to sulphate by other sulphur bacteria (see below) may prolong their dominance, but ultimately sulphate supplies will become limiting and methane-producing bacteria, the 'methanogens', become dominant. Methane formation represents the ultimate stage in this biological sequence: the familiar late autumn aquatic sediment, which evolves 'marsh gas' when disturbed, has passed through all these stages.

The hypothetical system just described was a fresh-water pond in which sulphate supplies became exhausted. In sea water or other sulphate-rich systems, sulphate does not become limiting and sulphate reduction is not normally superseded by methanogenesis. A transitional situation occurs in marine sediments, in which sulphate reduction is dominant in an upper zone but, at greater depths, methanogenesis takes over (see, for example, Hines & Buck, 1982; Jørgensen, 1982a, 1984).

Sulphates are common minerals in nearly all soils and waters, so sulphate reduction, once established, can continue for long periods provided organic matter is available. The sulphide formed is a chemical reducing agent, tending to preserve the anoxic environment necessary for activity of sulphate-reducing bacteria. Ecologically, zones of sulphate reduction are the foci of the eco-systems called 'sulfureta' (singular, sulfuretum; see also p. 4), in which other sulphur bacteria grow in association with the sulphate reducers. A schematic sulfuretum is shown in Fig. 15, in which coloured, photosynthetic sulphide-oxidizing bacteria grow in a zone whose depth is determined by the depth to which light penetrates. Such zones are common in deep stratified lakes and fjords. Above that zone, dissolved sulphide and dissolved air co-exist and aerobic sulphide-oxidizing bacteria (e.g. *Beggiatoa* in fresh waters, *Thiovulum* in the sea) may be found as well as thiobacilli. These organisms oxidize reduced sulphur back to sulphate and by thus re-cycling sulphur they tend to perpetuate the sulfuretum; by their autotrophic and photosynthetic CO_2 fixation they also regenerate organic matter and help to perpetuate the system. Since the sulphate reducers generate sulphide and the oxidizers fix CO_2, sulfureta can act as an anaerobic primary producing system, rather as plant photosynthesis is now the primary producing system for aerobic life on this planet. But sulfureta are limited by the amount available of either sulphide (for CO_2 fixation) or pre-formed organic matter (for H_2S formation). Though there is reason to believe that sulfureta were important ecosystems in the early history of this planet (sulphur deposits probably arose from gigantic sulfureta, see p. 138), they have not the universality of conventional photosynthesis, because this depends on the photolysis of water, not H_2S, and water is abundant today but H_2S is not. Nevertheless, there are many micro-environments, and some macro-environments, on this planet which provide the prerequisites of sulfureta: anaerobic conditions plus either organic matter (polluted soils and waters, decaying organic matter) or sulphide (artesian waters, putrefying protein, sulphur springs). These readily become populated by appropriate bacteria and, because of the reducing action of H_2S, have a considerable degree of ecological stability. They can become the basis of an eco-system supporting metazoa (Fenchel & Riedl, 1970). One of the most exciting examples of this kind is the sulfuretum discovered in the deep oceanic hydrothermal vents such as those described in the Galapagos Rift. Here H_2S-bearing hydrothermal water can support a complex biota including at least one possible symbiotic system (see Cavanaugh et al., 1981, and references therein). However, the primary source of H_2S is geothermal and though sulphate-reducing bacteria are present they are thought not to be the primary initiators of the eco-system.

Fig. 15. The sulfuretum in a pond. Decaying debris in the sediment provides organic matter for sulphate formation. At the lowest level to which light penetrates, a layer of anaerobic, photosynthetic sulphide-oxidizing bacteria develops. Above this, the water is aerobic, the sulphide concentration is low and aerobic sulphide oxidation can take place. Effectively, the top part of the sulphur cycle in Fig.1 is dominant in this eco-system.

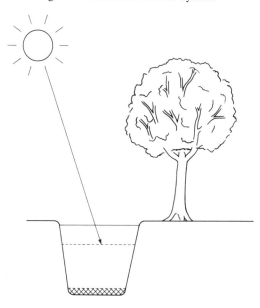

Sulfureta can readily be studied in the laboratory as dynamic entities, either enriched as mixed populations from nature (e.g. Bavendamm, 1924; Winogradsky, 1951; Suckow & Schwartz, 1963; Morgan & Lackey, 1965) or 'synthesized' by mixing pure strains of bacteria (e.g. Butlin & Postgate, 1954a, b; Matheron & Baulaigue, 1976; Biebl & Pfennig, 1978). They can also be studied in nature by conventional microbiological means as well as by measuring the conversion of ^{35}S-sulphate to sulphide and relating it to the transformations of ^{14}C-labelled organic matter or ^{14}C-labelled CO_2 (see Jørgensen, 1977, 1984).

Examples of sulfureta are found in salt marsh sediments (Howarth & Teal, 1979; Skyring, Oshrain & Wiebe, 1979; Nedwell & Abram, 1978, 1979), cyanobacterial mats (Jørgensen & Cohen, 1977; Bauld, Chambers & Skyring, 1979), sulphur springs (Butlin & Postgate, 1954a, b), marine sediments (Jørgensen, 1982a, b, 1984; Laanbroek & Veldkamp, 1982), peat (Sparling & Hennick, 1974) and many other environments where sulphate and organic matter co-exist in anoxic conditions. The references just cited are examples and are far from comprehensive, but they will provide a route into the now voluminous literature on sulfureta with special reference to sulphate-reducing bacteria. The ecology of sulfuretum in its various forms is a fascinating study which stretches far beyond the scope of this monograph; guides to it may also be found in the contributions to a symposium (Postgate & Kelly, 1982). However, a few points that bear particularly on the parts played by sulphate-reducing bacteria require brief discussion.

The role of acetate: For many years the sulfuretum had a paradox associated with it. Since most sulphate-reducing bacteria oxidize most organic matter only as far as the acetate level, it would be logical to expect acetate to accumulate in sulfureta, or else for some bacteria capable of anaerobic metabolism of acetate to be found (see, for example, Fenchel & Jørgensen, 1977). As I described in Chapter 1, numerous searches for such organisms failed to reveal either acetate-utilizing *Desulfovibrio* or other sulphide-tolerant acetate-utilizing anaerobes in samples from sulfureta. Yet all authorities agreed that acetate did not accumulate in sulfureta and that it disappeared slowly from enrichment cultures made from sulfureta. Recent microbiological studies by Professor N. Pfennig's group (see Pfennig & Widdel, 1982, and Chapter 2) have provided a satisfactory answer: the acetate-utilizing sulphate-reducing bacteria, including *Desulfobacter postgatei*, *Desulfotomaculum acetoxidans*, *Desulfovibrio baarsii* and *Desulfonema* species, can all consume acetate while reducing sulphate and doubtless do so in appropriate environments. Since sulphur is normally found on the more oxidized periphery of the sulfuretum, *Desulfuromonas acetoxidans* (Pfennig & Biebl, 1976) may also contribute to acetate consumption. Experiments with fluoracetate (to inhibit acetate utilization) and molybdate (to inhibit sulphate reduction) support the view that the two physiological classes of sulphate reducer (acetate-using and acetate-forming) co-exist in sediments (Banat, Lindström, Nedwell & Balba, 1981).

Sulphate reduction and methanogenesis: Sulphate-reducing bacteria have long been known to be incompatible with methanogens in ordinary laboratory culture, although they inhabit rather similar eco-systems and are both exacting anaerobes. Indeed, they seem in nature to have a somewhat competitive relationship, the methanogens 'taking over' when the sulphate-reducing bacteria have done what they can, yet being suppressed – but not eliminated – by sulphate reduction on any substantial scale. Cappenberg (1974*a*, *b*, *c*) studied their interrelation in Lake Vechten, Holland, and showed that their distribution was compatible with the sequence described earlier in this chapter: the methanogens were predominant beneath the sulphate reducers in the sediments. The inhibitor fluoroacetate inhibited methanogenesis and caused acetate to accumulate, suggesting that it could serve as a substrate for methanogenesis if too much sulphide did not accumulate. He also obtained evidence of lactacte accumulation, induced by fluorolactate, and regarded this as the natural substrate for sulphate reduction in such sediments. Thus a one-way commensalism appeared to exist in which sulphate reducers provided carbon for methanogenesis, and Cappenberg (1975*a*) traced labelled carbon from ^{14}C-labelled lactate emerging as ^{14}C-methane, probably via acetate. He successfully demonstrated such one-way commensalism in a chemostat (Cappenberg, 1975*b*). However, Abram & Nedwell (1978*b*) could not detect lactate accumulation and concluded that H_2 was the primary substrate for sulphate reduction in their sediments. The apparent incompatibility of sulphate reduction and methanogenesis was for many years perplexing. Acetate-utilizing sulphate-reducing bacteria were not known and, since acetate was known to be a good substrate for methane formation, at least one of the products of sulphate reduction (as then known) would be expected to favour methanogenesis. Many authorities attributed the incompatibility to sulphide toxicity: free sulphide was thought to be toxic to methanogens because, for example, sulphide added to CH_4-producing sewage sludge inhibited methanogenesis in proportion to its concentration (Butlin, Selwyn & Wakerley, 1956). However, more recent work has shown that sulphate reduction and methanogenesis are not incompatible at low sulphate concentrations (Oremland & Taylor, 1978) and that, far from being inhibited, some methanogens tolerate substantial sulphide concentrations (e.g. Ward & Olson, 1980) and some, indeed, are stimulated by sulphide (Mountfort & Asher, 1979). The work of Bryant, Campbell, Reddy & Crabill (1977) offered an explanation. They found that *Desulfovibrio vulgaris* and other types can associate with methanogens in the rumen's low sulphate concentration. They produce H_2 by sulphate-free metabolism which can then be used by methanogens for methane formation; the authors suggested that the incompatibility of methanogenesis and sulphate reduction might be a matter of H_2 scavenging, rather than sulphide poisoning, by the sulphate reducers. At high concentrations of sulphate, hydrogenase-containing sulphate-reducing bacteria would scavenge all available hydrogen for sulphate reduction, thus inhibiting methanogenesis.

This view has received widespread support. Abram & Nedwell (1978*a*) showed that sulphate reducers out-competed methanogens for exogenous H_2; Smith & Klug (1981), Nedwell & Banat (1982) and Sørensen, Christensen & Jørgensen (1981) have all shown that addition of molybdate to sulphate-reducing sediments, which inhibits sulphate reduction specifically, decreases the capacity of the sediment to take up exogenous H_2 and can lead to an increase in methanogenesis. Nedwell & Banat (1982) warned that experiments using molybdate are not rigid: it might have secondary effects not specific to sulphate reduction. However quantitative support has arisen from the studies of Kristjansson, Schönheit & Thauer (1982): the K_m values for H_2 uptake by methanogens lie between 6 and 15 µM whereas those of *Desulfovibrio* range from 1 to 3 µM. Clearly the desulfovibrios could out-compete the methanogens for H_2 in any normal eco-system, unless for some reason, such as sulphate deficiency, their capacity to oxidize hydrogen was limited.

Competition for H_2 accounts in part for the ecological separation of methanogenesis and sulphate reduction. However, more than half of the biogenic methane in nature is believed to originate from acetate. Thauer 1982; see also Schönheit, Kristjansson & Thauer, 1982) proposed that competition for acetate could also be expected: the K_m for acetate consumption by the acetate-utilizing methanogen *Methanosarcina barkeri* is about 3 mM, more than an order of magnitude greater than that of the acetate-utilizing sulphate reducer *Desulfobacter postgatei* (below 0.2 mM). In nature, the sulphate-reducer would out-compete the methanogen for acetate very effectively.

Mineralization of organic carbon: Until the discovery of the acetate-utilizing sulphate-reducing bacteria it seemed likely that sulphate reducers played a relatively minor part in the anaerobic mineralization of organic carbon, being concerned with the late stages along with fermentative clostridia but later than nitrate-reducing bacteria. They were thought perhaps to be involved in the penultimate stages of mineralization, with the methanogens responsible for the final stage. (Final because methane, being a gas, diffuses from anaerobic zones and becomes subject to aerobic mineralization by methanotrophic bacteria.) This situation may still obtain in fresh water and fresh-water sediments, where sulphate may readily become limiting, but there is now substantial evidence that sulphate reduction is a major mineralization process in sulphate-sufficient situations, particularly marine sediments. Jørgensen (1977) compared oxygen uptake and sulphate reduction as respective measures of oxygen-and sulphate-dependent mineralization in a temperate coastal marine sediment in Denmark and concluded that more than half of the mineralization was sulphate-dependent; he later (Jørgensen, 1982*b*) obtained comparable data for the sea bed, but the component of mineralization assignable to sulphate reduction decreased with depth and distance from the shore. Extrapolated to a global scale, sulphate reduction is responsible for a very substantial proportion of marine mineralization of organic carbon (see also Jørgensen, 1984).

Methane oxidation: A special case of mineralization applies to methane itself. As I mentioned earlier (pp. 45 and 68), there is no evidence from studies with authenticated pure cultures that methane oxidation coupled to sulphate reduction can occur. Nevertheless, sediments in which sulphate reduction is active often lie over sediments in which methane production is taking place, and several reports (collected by Jørgensen, 1982*b*) indicate that methane disappears in the sulphate zone. Ecologists are thus tempted to the view that methane can be a substrate for sulphate reduction, despite the elusiveness of the appropriate microbe (e.g. Barnes & Goldberg, 1976). Hydrolytic or carboxylative attack of methane followed by oxidation of the product, involving a syntrophy between a conventional sulphate reducer and another anaerobe as yet unidentified, could also be responsible, but one must not wholly disregard the possibility of mechanical or chemical errors in the detection of methane in complex systems such as sediments. Judgement on this matter must await the demonstration of a microbe or consortium capable of linking sulphate reduction to methane oxidation.

Nitrogen fixation: The agents of nitrogen fixation in the sea are largely pelagic cyanobacteria. These are not generally facultative anaerobes and are thus unlikely to be important in marine sediments. Evidence is now accumulating that sulphate-reducing bacteria may be quantitatively important in marine environments (e.g. Herbert, 1975). Dicker & Smith (1980) assigned a substantial proportion of the acetylene-reducing (nitrogen-fixing) capacity of a Delaware salt marsh population to sulphate-reducing bacteria. The absence of a correlation between acetylene reduction and counts of *Desulfovibrio* can probably be attributed to the use of unpoised counting media by both groups[1]. Nedwell & Azni bin Abdul Aziz (1980) used a different approach: they showed that molybdate in amounts sufficient to suppress sulphate reduction totally inhibited nitrogen fixation (assessed as acetylene reduction) in a temperate salt marsh and concluded that sulphate-reducing bacteria were the dominant diazotrophic anaerobes.

The economic effects of the environmental activities of sulphate-reducing bacteria, and of the establishment of sulfureta, are many-fold. Though they are intimately linked to the ecological effects described here, they will, for convenience, be discussed separately in the following chapter.

---

[1]See footnote to p. 30.

8

Economic Activities

Despite their inability to flourish in the normal aerobic environments on this planet, the sulphate-reducing bacteria impinge on mankind's economy in a surprisingly large number of ways. Detailed consideration of all known economic activities would be impossible within the confines of a monograph such as this; in earlier reviews I have attempted comprehensive cover (Postgate, 1960a, supplemented by parts of Postgate, 1965a, Le Gall & Postgate, 1973 and Postgate, 1982a). In this section their major economic functions will be described briefly with emphasis on the principles which underlie them; documentation will be minimal because the references in the reviews cited are reasonably comprehensive.

Pollution of water

Pollution is the introduction of excessive amounts of matter, usually organic matter, into an otherwise stable eco-system, with consequent destabilization of that system. The polluting matter is very frequently anthropogenic, but need not be so. Pollution with organic matter, sulphate or both often initiates a sequence of microbiological events which leads to large scale sulphide formation. Familiar examples of pollution of water leading to massive growth of sulphate-reducing bacteria occur in canals and harbours, where the ordinary detritus of human activity increases the biological oxygen demand of the water so that it becomes anaerobic and massive sulphate reduction ensues. The smell will be familiar to many summer visitors to Venice or Bruges. Marine or estuarine waters are particularly prone to such pollution because of the sulphate content of sea water. Baars (1930) reported that H_2S pollution of Dutch canals is enhanced by admission of sea water, a phenomenon he assigned to stimulation by salinity as well as sulphate. Paintwork on ships and nearby structures may be blackened by H_2S and Butlin (1949) attributed the traditional use of black paint for gondolas in Venice to common sense or this score. He described several comparable instances of nuisance caused by sulphate reduction in polluted waters.

Hydrogen sulphide evolved from polluted waters is primarily responsible for the 'nuisance' caused when domestic refuse is tipped on waterlogged sites, and it is one of the more disagreeable components of sewage. Ponds and waterlogged gravel or clay pits may become contaminated, causing complaints from nearby residents of bad smells (e.g. Bunker, 1942), tarnishing of paint and of copper or silver utensils. Domestic refuse tipped, sometimes by ill-advised local authorities, into clay or gravel pits that have been (or later become) flooded, is a common cause of nuisance, and is inclined to be particularly noisome towards the end of the summer. At this season the upper layers of water cool faster than the lower, and convection brings the sulphide-laden bottom waters to the surface. Despite the putative health-giving properties of water from sulphur springs, constant exposure to H_2S can be injurious to health. Deaths of sewerage workers from poisoning by H_2S of biological origin were recorded in Hungary by Vamos (1971); death was followed rapidly by *rigor mortis*. Occasionally such pollution leads to massive growth of photosynthetic sulphur bacteria with dramatic colour changes: 'lakes of blood', 'bloody seas' are often due to sulfureta in which blooms of the red *Chromatium* and *Thiopedia* are dominant (see, for example, Genovese, 1963; Butlin & Postgate, 1954*a*, *b*; Cabasso & Roussel, 1942). They have occasionally been considered miraculous (Jannasch, 1957).

Natural pollution of lakes can lead to death of fish (Fig. 16) and water plants. One of the most spectacular instances of water pollution occurs periodically at Walvis Bay, off South-West Africa (Copenhagen, 1934, 1953; Butlin, 1949) where a vast stretch of submarine mud, covering some 13 000 km^2, erupts every few years releasing sufficient H_2S to cause mass mortality of fish in the bay and gross atmospheric pollution over the seaside towns. In the winter of 1950–51 dead fish were strewn to a depth of several feet on the shore, bronze statues in the town were blackened and in February a new eruption 'blackened the face' of Swakopmund, a coastal town. The eruption continued, and in March three islands of brown mud, one 90 m long, rose up from 'a boiling sea' near a lighthouse at Pelican point, off Walvis Bay. The largest island persisted only one hour before subsiding into the sea; the smaller islands lasted several hours but had disappeared overnight according to Commander Copenhagen (personal communication). Biogenic H_2 causes death of fish in the Sea of Azov in winter, according to Tolokonnikova (1977).

Dr R. Vamos (see Le Gall & Postgate, 1973) has described the 'poisonous dawn fog' believed by Italian and Hungarian peasants to kill aquatic birds; it is bacterially produced H_2S formed during the decay of algal blooms and it can kill fish at a rate of some 25 tonnes in 3 hours (e.g. Fig. 16). Suspended lime in the water can have a protective action and the toxicity to fish may result less from the H_2S than from acid produced by the oxidation of biogenic FeS, according to Dr Vamos and his colleagues.

Fig. 16. Fish killed by sulphide pollution. Dead carp killed by biogenic H_2S as a result of pollution in Lake Palic, Yugoslavia, in 1971 (courtesy of Dr R. Vamos).

Pollution of water   127

Moderate pollution of the water may permit thermophilic sulphate-reducing bacteria to grow in domestic hot water and central heating systems as well as closed industrial water cooling systems. In such circumstances corrosion (see below) often results. They can also grow in the aqueous phase of oil and petrol storage systems, in the water bottoms of gas storage systems, in stagnant cutting emulsions and so on; their effects in these contexts are diverse and are discussed later in this chapter. Wood (1961) described a curious variant of water pollution: a situation in which the iron content of drinking water was augmented sufficiently to cause 'rusty water', because sulphate-reducing bacteria were corroding the main. This phenomenon occurs only with deep well water, which is almost anaerobic; it can be cured by aeration (see Windle-Taylor, 1962).

The economic and social importance of such phenomena can be considerable, yet, due to the difficulties of managing large volumes of water, there is no straightforward solution to the problem. In general the most promising approach is inhibition of the bacterial growth and metabolism, though the possibilities of biological control have not been exhaustively studied. Sterilization is almost always out of the question. The treatment depends on the nature of the primary pollutant, as well as on the site and environment of the water polluted, and because of the high resistance of sulphate-reducing bacteria to most inhibitors the number of available remedies is small. Available treatments fall into three major classes (see also Postgate, 1960a).

Avoidance or confinement of pollution

The best and cheapest inhibitor of bacterial sulphate reduction is air. Provided some O_2 remains in solution the E_h will be too positive for growth of these bacteria and, even in mildly polluted waters, they will cause no trouble. Sometimes, however, gross organic pollution is inevitable, in which case it may be possible to confine the pollution by limiting the volume of water polluted at any given time. As an example, Knolles (1952) used such a procedure for tipping urban domestic refuse into waterlogged gravel pits, thus reclaiming building land as well as disposing of refuse. He divided the pits into lagoons by tipping walls of clinker, the size of the lagoons being such that they could either be filled completely with refuse before massive growth of sulphate-reducing bacteria had time to occur, or controlled by acidification (below) should this prove necessary. Deliberate aeration can sometimes be used to protect drinking water which is to be piped in circumstances in which conditions may readily become anaerobic, but, in general, injection of air into water masses is an expensive operation.

Acidification

Bunker (1942) described an instance in which a pond (36×10^6 litres), into which waste from a firm manufacturing edible fats was tipped, became badly polluted and caused complaints from local residents. Some 42 tonnes of 76% H_2SO_4 (v/v) were added, which brought the pH value down to 3, and nuisance ceased. Such treatment is rarely feasible unless the sulphuric acid is available cheaply as an industrial waste; acidification with HCl has been used in the oil industry to control these bacteria (Anderson, 1957).

Biological control

The possibility of using sulphide-oxidizing bacteria for the control of H_2S pollution is an attractive idea (e.g. Butlin & Postgate, 1954b) which would mimic the situation in natural sulfureta. However, when the H_2S concentration reaches a level likely to cause nuisance, the consequent blooms of phototrophic sulphur bacteria do not cope and H_2S escapes into the aerobic zone. This may arise from toxic effects of relatively high H_2S concentrations on *Chromatium* and *Thiobacillus* species; the isolation of highly sulphide-tolerant sulphide oxidizers such as *Thiomicrospira pelphila* (Kuenen & Veldkamp, 1972) suggests that biological control might profitably be re-investigated.

Chemical inhibition

Appendix 2 and the discussion in Chapter 3 (p. 47) made it clear that 'chemotherapy' of sulphate reduction is an unpromising prospect. However, a few inhibitors are available which are cheap enough to use in confined environments. An example is 'dye 914' (see Appendix 2) which is an acridine used by Rogers (1940, 1945) to preserve stored submarine cables and to prevent sulphate reduction in gas holder waters. Some quaternary detergents are available as by-products of the meat industry and have been used to control problems encountered by the oil industry (see Beerstecher, 1954). The high resistance to quaternaries sometimes associated with *Desulfovibrio salexigens* can cause difficulties. Nitrates have a transient inhibitory effect and, in such environments as rice fields, are agriculturally beneficial. Chromate solutions are sometimes available cheaply as waste from chromium plating industries and can be very effective indeed as a prophylactic in potentially dangerous situations. Apse, Morwood & Wood (1955) and Drummond & Postgate (1955) reported a case in which chromate was used to control pollution of a waterlogged clay pit into which domestic refuse was being tipped. It did not influence the growth of aerobes associated with ordinary pollution but, while chromate remained detectable in the water (down to 0.2 μg, CrO_3/ml), the population of sulphate-reducing bacteria remained negligibly small and no nuisance occurred. Further discussion and documentation of control procedures may be found in Beerstecher (1954) and the reviews cited at the opening of this chapter.

Release of phosphate in water

Hayes & Coffin (1951) suggested that biological production of phytoplankton, as food for fish, depended on release of phosphate from basic ferric phosphates in fresh-water bottom muds by sulphate-reducing bacteria: conversion to FeS released soluble phosphate, a reaction demonstrated by Sperber (1958).

Pollution of soil and sand

Marine mud and sand, in harbours and estuaries where organic pollution levels are high, have a property familiar to most tourists: its surface is sandy brown but, when dug into, the newly exposed sand is black. The black colour disappears in a matter of minutes on exposure to air. The explanation (Ellis, 1932; Bunker, 1936) is that microbial sulphide formation leads to precipitation of black FeS and that this autoxidizes to brown products (ferric hydroxide and sulphate) in air. When the tide returns, anaerobic conditions will be favoured and the sand will blacken. Sulphate-reducing bacteria are the major contributors to this cycle although, as Gunkel & Oppenheimer (1963) pointed out, a contribution is made by other bacteria decomposing organic S compounds. Apart from having a negative effect on tourism, for the sand smells lightly of H_2S and its appearance is unattractive, such reversible blackening of sand is probably of little economic importance. However, the phenomenon serves as a warning that metal and stone installations in such environments are threatened by microbiological corrosion (below).

Soils wholly deficient in sulphate-reducing bacteria are rare but the bacteria are inactive if soils are aerobic. Soils showing pronounced blackening or clays showing 'gleying' are normally anaerobic and often heavily infected. They present a threat of corrosion and can also cause 'nuisance'; an example is a case of 'blackened and stinking soil' described by Postgate (1960a): the smell, which was annoying local residents and darkening paint and metalwork, was traced to buried builders' refuse and organic debris supporting exuberant activity of sulphate-reducing bacteria.

On the credit side, certain soils infected by sulphate-reducing bacteria are believed to have medicinal properties. 'Ripening' of the hot medicinal muds at Piestany-spa, Czechoslovakia, is associated with the activities of thermophilic sulphate-reducing bacteria (Starka, 1951).

Rice paddies are environments which are particularly prone to H_2S pollution. The plants themselves are relatively resistant to sulphide, but the H_2S concentration may nevertheless reach damaging proportions and the crop fail. Wilting for this reason tends to occur in the autumn and may lead to considerable crop losses (for references, see Le Gall & Postgate, 1973). Nitrates, which inhibit sulphate reduction and increase fertility, have been used to control such damage and Takai & Kamura (1966) recorded that control of the redox potential in paddy soils has increased yields by a third. Sulphide in rice paddies is not wholly damaging, however: it kills parasitic nematodes, so sulphate reduction can actually be exploited for protection of the rice crop (Jacq & Fortuner, 1980). Soils rich in organic matter, notably the black alkaline chernozem, can become toxic due to sulphate reduction, and growth of sulphate reducers in the rhizosphere of crops other than rice can, in some soils, kill the plant or stunt its growth. Examples of such damage include death of citrus plants (Ford, 1965), legumes (Dommergues et al., 1969) and maize (Jacq & Dommergues, 1971). Germinating seedlings are susceptible; high insolation and high sulphate concentrations seem to favour such toxicity (Jacq & Dommergues, 1970).

Purification of wastes

Most of the topics described so far illustrate the negative effects of sulphate-reducing bacteria on man's economy, but possibilities for biotechnological exploitation of these bacteria also exist. For example, sewage and several types of industrial waste liquors are treated by microbiological purification processes before being released into the environment. Oxygen is the usual electron acceptor for such processes (e.g. the activated sludge process), though the methane fermentation is used in the treatment of settled sewage sludge. McKinney & Conway (1957) discussed sulphate as a possible terminal oxidant for anaerobic biological waste treatment and Pipes (1960) developed a potentially practicable process using first an artificial waste (hair + starch + filter paper) and later activated sludge. Domka, Gasiowek & Klemm (1977) surveyed a variety of municipal wastes (sewage; wastes from yeast, dairy and sugar plants) for processing by way of bacterial sulphate reduction.

The sulphite-containing waste liquor of the paper industry presents a disposal problem because sulphite inhibits the normal methane fermentation (Noordam-Goedewagen, Manten & Muller, 1949). Bannink & Muller (1952) passed waste liquors from the digestion of straw with sulphite through a column impregnated with sulphate-reducing bacteria and removed sufficient sulphite from the liquor for it to undergo an ordinary methane fermentation. Freke & Tate (1961) showed that, in certain conditions, magnetic iron sulphide was formed during bacterial sulphate reduction and the Fe could then be removed with a magnet. Unfortunately, conditions for reproducible formation of the magnetic product were not established or a process for the removal of iron from ferruginous waters could have been developed. Sulphuric acid in mine waters can present a serious disposal problem; Tuttle and his colleagues (Tuttle, Randles & Dugan, 1968; Tuttle, Dugan & Randles 1969; Tuttle et al., 1969) described a system in which acid mine water flowed through a porous dam of wood-dust, within which a consortium of cellulolytic and sulphate-reducing bacteria reduced sulphate and generated sufficient alkalinity to render the effluent acceptable. Iron, sulphate and acidity were thus removed in one treatment.

Several processes for effluent treatment have arisen in the context of the microbiological production of reduced sulphur; some will be discussed later (p. 140). Distillery slop has been subjected to sulphate-dependent treatment in India (Basu & Ghose, 1961; Ghose & Basu, 1961; Ghose, Mukkerjee & Basu, 1964) and wastes from a distillery and citric acid plant were so treated in Czechoslovakia (see Barta, 1964). The process developed by Butlin and his colleagues (p. 140) made use of digested sewage sludge as the primary reductant for sulphate; it had the advantage that the resulting sulphide-containing sewage sludge had improved settling qualities which made it considerably more amenable for disposal. The reason is that, with normal digested sewage sludge, settling or 'dewatering' is interfered with by residual methane production in the sludge, so that the final 'settled' product contains as much as 97% water. Sulphide-fermented sludge does not produce gas and the water content can rapidly be lowered

132 Economic activities

in settling tanks to 92%. If the digested sludge has to be transported any distance for final disposal this improved settling brings considerable savings because the problem of moving great quantities of water is avoided: up to four times as much water may be removed after settlement from sulphide digestion (*Chemistry Research*, 1957).

However, methane is an economically valuable product of sewage purification by anaerobic digestion. The economic value of sulphate digestion for the purification of wastes thus depends on the following three questions.

1. Does the advantage in settling properties of the final sludge outweigh the loss of methane, which is a useful source of energy?

2. Is its applicability to strong or otherwise unacceptable wastes, such as distillery slop or sulphite-waste liquor, worth the capital cost of remodelling plant and arranging for extraction of H_2S?

3. Does a sufficient market exist for the H_2S obtained as a by-product to render the process economic?

Formation of sulphate-deficient environments

Given a high ratio of metabolizable carbon to sulphate, sulphate-reducing bacteria can deplete an environment of sulphate almost completely. Some lakes in East Africa are markedly sulphate-deficient and S in fact limits biological productivity; some soils in that locality are also S-deficient. Beauchamp (1953) suggested that massive sulphate reduction might be responsible but this now seems improbable. Hesse (1956) reported that the presence of sulphate in the lake mud, and its oxidizing redox potential, was inconsistent with significant sulphate reduction (see also Sugawara, Koyama & Kozawa, 1953; Koyama & Sugawara, 1953).

Corrosion of metals and stonework

Underground corrosion of iron or steel gas or water pipes was costing the USA 0.5 to 2×10^9 dollars a year some 25 years ago (Greathouse & Wessel, 1954) and is the best known economic disaster engendered by sulphate-reducing bacteria. The subject has been surveyed by many authorities including Starkey & Wight (1945), Starkey (1958), Postgate (1963*c*), Booth (1964), Miller (1971, 1981), Iverson (1972, 1974), Crombie, Moodie & Thomas (1980) and an extensive technical literature exists, so the treatment here will be brief. For detailed documentation the surveys just cited should be consulted.

Anaerobic corrosion of iron of the kind brought about by sulphate-reducing bacteria has three main characteristics:

1. It is restricted to anaerobic environments such as clay or waterlogged soils. When freshly exposed, the soil in the neighbourhood of the corroding metal is blackened with iron sulphide and often smells of H_2S. Polluted harbours and areas beneath marine encrustations can provide appropriate anaerobic environments; 'tubercules' of mixed iron oxides which develop on the inside of water pipes can harbour sulphate-reducing bacteria and permit corrosion from inside the pipe. Water beneath spirit in petroleum storage tanks, or gas holder waters, also provides environments favouring anaerobic corrosion of any iron which may be present, as do many enclosed water-cycling systems. Iron objects in areas subject to intermittent or partial anaerobic conditions often corrode faster than those in wholly anaerobic conditions.

2. The corroded metal tends to be pitted rather than evenly corroded, indicating that the corrosion is not of a self-stifling character. Breakdown of iron pipes caused by bacterial corrosion thus tends to take the form of local perforation rather than wholesale deterioration.

3. The metal at the point of corrosion is characteristically graphitized: the metallic iron is entirely removed but the graphite skeleton of the pipe often retains its original form (see Fig. 17).

The mechanism of anaerobic corrosion by sulphate-reducing bacteria is still somewhat controversial. Sulphide is in itself corrosive, and corrosion of metals by sulphide-producing bacteria other than the sulphate-reducing bacteria has been recorded, but the distinctive feature of corrosion by *Desulfovibrio* and *Desulfotomaculum* is cathodic depolarization of iron by the bacterial hydrogenase. The theory of cathodic depolarization was expounded in 1934 by von Wolzogen Kühr & van der Vlugt (1934) and has provoked much research and argument since then. In principle it is simple. A clean ferrous surface reacts with water and rapidly becomes covered with a film of hydrogen which depolarizes the surface and prevents further reaction. This reaction can be expressed as

$$2H_2O \longrightarrow 2H^+ + 2OH^-$$
$$Fe^{2+} + 2OH^- \longrightarrow FeO + H_2O$$
$$\searrow Fe_2O_3, \text{etc.}$$

$$+ \longrightarrow +$$

$$Fe$$
$$|$$
$$Fe-Fe-Fe-Fe$$

$$H-H$$
$$\vdots \quad \vdots$$
$$Fe-Fe-Fe-Fe$$

where the iron phase is solid and the superficial hydrogen derivative is capable of behaving either as hydride or as adsorbed dihydrogen molecules. In rusting, dissolved O_2 slowly removes the cathodic film of hydrogen, permitting the next steps: repetition of the process and conversion of the metallic iron to ferrous hydroxide and thence to the mixed ferrous and ferric oxides of rust. In anaerobic corrosion, the sulphate-reducing bacteria can remove the hydrogen film through their enzyme hydrogenase with net formation of both iron hydroxides and iron sulphide in the corrosion product (Fig. 17). The ratio of hydroxide to sulphide is influenced by the extent of organic pollution of the environment: if much organic matter is present (as in many laboratory simulations of underground corrosion) the corrosion product will be entirely sulphide because the hydroxide reacts with free H_2S.

The argument about the mechanism of anaerobic bacterial corrosion centres not on the involvement of sulphate-reducing bacteria; of that there is no doubt. The debated question is the extent to which cathodic depolarization contributes to the mechanism; for details the reviews cited

Fig. 17. Anaerobic corrosion of a spun steel water pipe. The corroded pipe has been cut to show characteristic corrosion beneath tubercular deposits on the inner surface. Corrosion tends to be localized; some attack has been initiated from the outer surface. The darkening is 'graphitization', loss of iron, leaving the carbon skeleton of the steel.

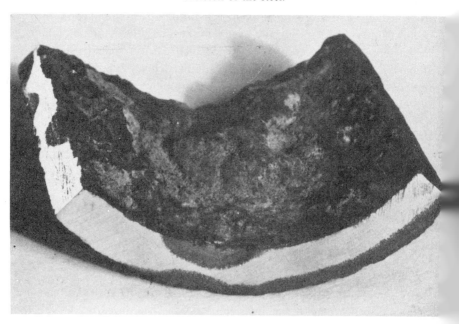

at the beginning of this section should be consulted. Cathodic depolarization undoubtedly takes place: it is readily demonstrated in the laboratory and correlations between corrosion rate and hydrogenase content of the strain can be shown although they are not necessarily quantitative (Booth *et al.*, 1965; see also Le Gall & Postgate, 1973). However, it is clearly only one of several important processes which lead to underground corrosion of ferrous metals. Other processes also contribute in practice and may become dominant in special circumstances. Indeed, Starkey (1958) expounded the view that cathodic depolarization makes but a minor contribution to underground corrosion. Examples of alternative or supplementary processes include the following. Direct attack on iron by H_2S proceeds at an appreciable rate and, where air has partial access to the system (e.g. the surround of a tubercle harbouring sulphate-reducing bacteria in a water pipe, or an iron installation in the tidal zone of a sea shore), oxido-reduction cells ('differential aeration cells') may be set up on the metal surface which corrode the metal at their cathodic sites. Oxidation of sulphide to free sulphur, itself highly corrosive, is catalysed by iron and may also occur. Studies with organisms grown in non-sulphate-reducing conditions showed elegantly that solid FeS itself can be corrosive (Booth, Elford & Wakerley, 1968). The extent to which the types of corrosive reaction dominate in a given environment depends on a variety of factors – the nature of the metal surface, the presence or absence of dissolved iron and/or organic matter capable of chelating iron in the surrounding water, whether the strain of bacteria tends to form a film on the metal itself, whether the iron sulphide itself forms a film (which can be protective), whether other ions such as Na^+, Cl^- (usually synergistic) or phosphate (usually somewhat protective) are present. Certainly the interface between a highly anaerobic and a partly aerated soil is a highly corrosive zone (see Starkey, 1958).

Despite the obvious economic importance of anaerobic corrosion, useful estimates of its economic cost are difficult to make. It must clearly be very high: replacement of small corroded gas or water service mains can cost £20 per metre, rising to £300–400 per metre for larger mains (Wakerley, 1979). An estimate of £20 × 10^6 as the total cost of such corrosion to the UK economy is over 25 years old (Vernon, 1957) but it appears to be the most recent, albeit parochial, estimate available.

The problem of whether a particular soil will be 'aggressive' towards buried iron pipes or other installations is an important practical one. Well-aerated soils are not, but any soil which is waterlogged and likely to become anaerobic, particularly intermittently, can be aggressive. Counts of numbers of sulphate-reducing bacteria are of little value, even when performed with good media (see Chapter 3), because they do not distinguish between active and dormant desulfovibrios. Starkey & Wight (1945) recommended E_h as a guide and proposed the following scale:

E_h range (mV)	Corrosiveness
Below +100	Severe
100 to 200	Moderate
200 to 400	Slight
Above 400	Non-corrosive

Booth and his colleagues (Booth, Cooper, Cooper & Wakerley, 1967; Booth, Cooper & Cooper, 1967; Booth, Cooper & Tiller, 1967) examined this question and, agreeing that bacterial counts were of little value, recommended a combination of redox potential and soil resistivity as a guide to aggressiveness. Yet the work of Starkey and Hubble (see Starkey, 1958) would lead one to expect a fluctuating E_h to provide the most aggressive environment of all.

Though damage in the field to buried iron and steel pipes and installations is the commonest manifestation of anaerobic bacterial corrosion, one should not forget that thermophilic desulfotomacula can cause corrosion in hot environments. Two examples I know of are corrosion in 1946 of an electrical transformer tank buried in London clay, where the temperature was 60–80 °C for long periods, and corrosion of a metal tank carrying hot molasses at a sugar refinery. In both instances the corrosion product contained FeS and *Desulfotomaculum nigrificans* was present. Metals such as zinc, aluminium and copper can also be corroded by these bacteria, the mechanism presumably being direct attack by H_2S, but cases of the kind seem rather rare. Non-sulphate-reducing bacteria which can generate H_2S from organic sulphur compounds may contribute to corrosion by direct sulphide attack in appropriate environments (e.g. Brisou & De Rautlin de la Roy, 1965).

The prevention of anaerobic bacteria corrosion is a vitally important practical question, and again the reader must be referred to the extensive technical literature for details. In principle four methods are available:

1. Surround the pipe or other object with a zone of non-aggressive soil (e.g. 20–25 cm of sand, chalk or gravel). This will ensure aerobic conditions, incompatible with bacterial sulphate reduction, provided the surround does not become waterlogged.

Corrosion of metals and stonework

2. Protect by coating the metal. Pipes can be protected by coating them with bitumen, rubber derivatives, plastics, waxes or concrete, so physically preventing access of the bacteria to the metal surface, a method which works if the coating remains undamaged during laying. Fig. 18 illustrates an oil pipe being wrapped with fibreglass and bituminous material in the Middle East.

3. Use chemical inhibitors for closed systems (such as a central heating system).

4. One of the most reliable processes is called 'cathodic protection'. This method involves the generation of a direct current either from an electrical mains or by connecting the iron pipe to a 'sacrificial' magnesium anode, which imposes a negative potential on the iron so that it is constantly polarized despite the activities of the bacteria.

Sulphate-reducing bacteria can contribute to a second type of corrosion, or deterioration, in which stone and concrete work can become corroded. The mechanism is different: if H_2S from a polluted soil or water diffuses to an aerobic zone, thiobacilli oxidize it to sulphuric acid which is itself corrosive. The most common source of such H_2S is biological sulphate

Fig. 18. Wrapping an oil pipe against anaerobic corrosion. An iron pipe in the Middle East being wrapped mechanically with fibreglass (two layers) followed by a bituminous skin to prevent access of sulphate-reducing bacteria (approximately 1950; courtesy of the late K. R. Butlin).

reduction, but H_2S in effluents from the chemical industry can be the source. The classical instances of this class of corrosion occur in concrete sewerage pipes, where H_2S from the sewage is oxidized on the roof of the pipe to sulphuric acid and ultimately the roof collapses. Corrosion of stone statues set in polluted soils, and of temples in tropical areas, such as Angkor Wat in Cambodia, originates in a comparable way: water bearing biologically produced H_2S seeps up the stonework by capillarity and becomes oxidized by thiobacilli to corrosive sulphuric acid. The technical description of this phenomenon as a 'disease' of stonework (Pochon, Coppier & Tchan, 1951) is not entirely inappropriate. Acid corrosion of this kind has been discussed by Krumbein & Pochon (1964) and Purkiss (1971); Iverson (1972) quoted an example of corrosion of buried copper piping by biogenic sulphuric acid.

Formation of mineral deposits

Among the more constructive economic activities of the sulphate-reducing bacteria has been the formation of several important classes of mineral deposit. These will be discussed under separate headings.

Native sulphur

Over 90% of the world's supplies of natural sulphur are in Texas and Lousiana. The sulphur isotope ratio indicates that the deposits are biogenic and they are believed to have been formed as a result of huge sulfureta developing during the evaporation of enclosed seas in the Permian or Jurassic era (about 2×10^8 years ago). Not all authorities are agreed on the detailed mechanism or period, but the role of sulphate-reducing bacteria is not disputed. Biogenic sulphur is found in the USSR, Sicily, Iceland and New Zealand. Domka & Gasiowek (1975) attributed the formation of sulphur deposits in the Carpathians to a primary bacterial reduction of gypsum. The involvement of these bacteria in sulphur formation was discussed by Ivanov (1964) and Trudinger (1976).

In Libya certain lakes exist where contemporary biogenesis of sulphur may be observed. One of them, Ain-ez-Zauia (Fig. 19), yielded some 100 tonnes of crude sulphur a year according to the local inhabitants; three lakes in the neighbourhood were exploited, and they were distinguished from a fourth, which yielded little sulphur, by the presence of a red gelatinous 'carpet' beneath the shallower reaches of the lakes (Butlin & Postgate, 1954a). The water of Ain-ez-Zauia was rich in dissolved H_2S

Formation of mineral deposits 139

(up to 100 mg/litre), saturated with calcium sulphate and containing about 2% NaCl (pH 7.4); an artesian origin was indicated by the fact that its temperature was about 30 °C. Little organic material was present, but it was heavily infected with *Desulfovibrio* and the red gelatinous material consisted of a zoogloeal mass of coloured photosynthetic sulphide-oxidizing bacteria (*Chromatium* and *Chlorobium* spp.) with a certain amount of occluded gas (methane?) which was not identified. The red gelatinous material could be 'grown' in laboratory conditions in illuminated tanks, and its growth was associated with sulphur formation; using pure cultures isolated from the lakes, Butlin & Postgate (1954a) showed that the combinations *Desulfovibrio* + *Chlorobium* and *Desulfovibrio* + *Chromatium* formed sulphur from sulphate when illuminated in appropriate media. The former combination was most probably responsible for sulphur formation in the lakes examined, since the sulphur was deposited outside the

Fig. 19. View of Ain-ez-Zauia, Libya. A saline, sulphur-producing lake in Libya, fed from an artesian source at 30°C. The milkiness of the water is due to suspended sulphur; the edge is encrusted with sodium chloride and mixed sulphate and carbonate of calcium. *Desulfovibrio desulfuricans* strain El Agheila Z (Fig. 3) originated here.

bacterial cells. A sample of sulphur from Ain-ez-Zauia showed its biological origin because it was enriched in the light sulphur isotope compared to calcium sulphate from the same site (see Chapter 5, p. 89). Butlin & Postgate (1954a) considered that the photosynthetic sulphur bacteria provided organic material for sulphate reduction by CO_2 fixation; the sulphate-reducing bacteria in return provided sulphide for photosynthesis by the coloured bacteria. Consequently the net process taking place was reduction of sulphate to sulphur at the expense of solar energy.

The lakes in North Africa, like sulphur springs anywhere, are examples of a natural sulfuretum (see p. 118). They are occasionally exploited but their productivity is trivial compared with the sulphur requirements of an industrial country. In 1962, 5×10^6 tonnes of sulphur were marketed in the USA, 1.3×10^6 tonnes in Mexico and France. The world's supplies of native sulphur are limited and, though new deposits were discovered in the 1960s, the rate of exhaustion of old deposits was exceeding the rate of discovery of new ones again during the 1970s. The major industrial demand for native sulphur is for the production of sulphuric acid, which has an enormous number of industrial uses; the element sulphur itself has relatively few uses. Pyrites and anhydrite can be used as sources of sulphuric acid, but sulphur is a very convenient source. Therefore the possibility of exploiting processes analogous to the natural biogenesis of sulphur has attracted a certain amount of research. Indeed, the microbiological production of sulphur reached pilot plant stages as a biotechnological process in the early 1960s. From the point of view of industry, chemical conversion of sulphide to free sulphur or sulphuric acid is relatively easy; the difficult, power-consuming step is the reduction of the sulphate ion. Therefore microbiological production of sulphide is potentially as useful as microbiological production of sulphur.

A hydrogen donor is obviously needed for sulphate reduction; in a microbiological process this may be gaseous H_2, the products of other autotrophic bacteria as in the Libyan lakes, or a waste organic material. In practice, organic wastes are the most plausible exploitable resources and sewage is virtually the only material which is produced by industrial societies on a sufficient scale to be economically interesting. Butlin, Selwyn & Wakerley (1956, see also Knivett, 1960) described a process for producing sulphide using sewage sludge 'activated' to produce H_2S by repeated subculture in the presence of sulphate. In a semi-continuous process based on routine sewage works practice they were able to obtain a gas mixture of 60% CH_4, 30% CO_2 and 5–10% H_2S, using methane from a separate fermentation to remove H_2S from the fermentor in which the sulphide was being formed. Production of such gas mixtures from a single fermentation would not be practicable since the methane fermentation is inhibited by too much sulphate reduction (p. 120); they recorded preliminary attempts to obtain tolerant methane-producing sludges which might make a single fermentation to yield both methane and H_2S more feasible. They concluded: 'The microbial production of sulphide from sewage sludge, while technically feasible, would entail using raw sludge which could otherwise be used for methane production. In the event of another shortage of sulphur, diversion of part of the available sludge for the production of sulphur rather than methane might well be justified.' The context referred to the United Kingdom, but applies to any country having centralized sewage disposal and lacking indigenous resources of reduced sulphur.

Butlin et al. (1960) later described the effects of altered temperature, gas flows and retention times on the sulphide fermentation of sewage sludge, and obtained maximum yields of about 1 g sulphide per 100 ml sludge fermented. An unexpected bonus from this research was mentioned earlier (p. 131): the settling properties of sulphide-fermented sewage were superior to those of sewage subjected to a methane fermentation. Burgess & Wood (1961) described large-scale experiments conducted at a London sewage works in collaboration with Butlin and his colleagues and pointed out certain disposal and corrosion problems that the sulphide fermentation entailed.

It is possible to calculate from Butlin's data that a large sewage works processing about 6000 tonnes of sludge a day could produce about 20 000 tonnes of sulphide a year. About 95% of this could be removed by 'stripping' with sewage gas (or another gas mixture containing about 30% CO_2); the effluent would contain about 50 mg H_2S/litre which would present the disposal problem discussed by Burgess & Wood (1961); the stripping gas would carry 5 to 10% (v/v) H_2S, which could readily be recovered as sulphur or sulphur oxides by existing industrial techniques and the residual gas re-cycled. The late K. R. Butlin (personal communication) estimated that the amount of sewage sludge processed in Britain would be equivalent to 10^6 tonnes of sulphur/year; in 1950 Britain's requirement was 2.5×10^5 tonnes/year. Since less than 10% of Britain's sewage could in practice be so processed, and the demand for sulphur increases annually, it is clear that microbial sulphur production could only be a supplementary source of material for a highly industrialized country.

The situation typifies the problems of microbial production of bulk chemicals. Generally speaking, microbiological processes are not readily adaptable to the production of simple chemicals required in large amounts ('heavy' chemicals). Sulphur is such a chemical. It is a fossil reserve of chemical energy analogous to coal, deposited by biological action in past geological eras. Like coal and oil, the world's native sulphur supplies must in due course become exhausted, and industry is already being obliged to exploit other forms of reduced sulphur (e.g. pyrites) or to reduce sulphate chemically. Relatively little of the sulphur consumed by industry appears in the industry's product; nearly all of it is consumed as sulphuric acid which becomes diluted and eventually finds its way to the sea (the main exception to this generalization is provided by the relatively small amount of sulphur incorporated into vulcanized rubber). Thus the industrialization of this planet has led to a net loss of fossil reserves of concentrated, reduced sulphur and their dispersal as dilute sulphates. This is a familiar pattern for the treatment of the world's fossil energy reserves during the present century, and in the long run the trend must be reversed by the expenditure of some new forms of energy (atomic, tidal or solar). The world's supplies of coal and oil are expected, barring war, to last until atomic energy becomes more economically available. The interesting situation regarding the world's sulphur resources is that they may well not last that long. As with copper, a situation has already been reached once in which the rate of discovery of new deposits was less than the rate of exhaustion of old ones, and there is every reason to expect the situation to recur with increasing frequency. At no time would microbial sulphide production justify itself economically as a biotechnological process for producing reduced sulphur (except perhaps in the artificial economic circumstances engendered by a war or a catastrophic energy shortage). But, for industrialized countries that are not primary producers of sulphur, microbiological sulphide production should be taken seriously as an important part of long-range economic strategy to guard against the next shortage, and could justify itself in the interval by its superiority in certain respects (fat removal, sludge settlement, applicability to strong effluents) to methane fermentation.

Formation of metal sulphide ores

As a consequence of bacterial activity in a sulfuretum, heavy metals tend to precipitate as sulphides and a diffusion gradient of heavy metal ions is formed which favours accumulation of the metal in the sulfuretum. Sulphate-reducing bacteria have been believed to have been involved in the genesis of metal sulphide ores (e.g. those of lead and zinc) since the early

Formation of mineral deposits

decades of this century (e.g. Siebenthal, 1915; Bastin, 1926; Bunker, 1936). The topic was discussed by Silverman & Ehrlich (1964) and Trudinger (1976). ZoBell (1946) mentioned that they might be involved in the deposition of copper sulphide ores; Miller (1950) grew these bacteria in the presence of carbonates or oxides of Pb, Zn, Sb, Bi, Cd, Co and Ni and considered that they might be responsible for the deposition of several types of metal sulphide ore. Cu as the basic carbonate showed some toxicity, a point taken up by Booth & Mercer (1963), who concluded that sulphate-reducing bacteria play no significant part in the genesis of copper sulphide ores. However, a well established sulfuretum could precipitate a lot of copper without the copper ion reaching an inhibitory concentration (Suckow & Schwartz, 1963), so a continuous supply of Cu^{2+} at low concentrations could yield copper sulphide deposits (see Temple & Le Roux, 1964a). Suckow & Schwartz (1963) believed that sulphate-reducing bacteria contributed to the formation of the copper sulphide shales at Zechstein in Germany, and set up sulfureta to examine the effects of added copper or iron salts to them.

Most deposits of iron sulphide ores are believed to have been formed in

Fig. 20. Carbonate encrustations around water plants killed by biogenic H_2S in a polluted lake in Hungary (courtesy of Dr R. Vamos).

this manner and isotope ratio measurements have confirmed this view for some ferrous deposits, notably pyrites, which are formed by way of an iron hydroxysulphide, hydrotroilite. Trudinger (1976, 1982) reviewed the role of these bacteria in ore formation; the distribution of S isotopes suggests that sulphide deposits of Cu, Ag, Pb, Zn and U could also be biogenic but Trudinger (1982), in contrast to Temple (1964), concluded that the biological contributions to most of these ores is, at best, indirect.

A special case is presented by the formation of pyritized fossils. It seems likely that the slow putrefaction of the organism leads to micro-sulfureta, causing accretion of hydrotroilite in zones following closely the structure of the original creature. Subsequently transition to pyrites takes place. It is remarkable how faithfully the original structure can be reproduced, given that biogenic precipitates of metal sulphides can be discontinuous (Temple & Le Roux, 1964b).

Formation of soda deposits

Light metals such as sodium or calcium do not form insoluble sulphides, but the sulphide dissociates to give free H_2S and OH^-. Since CO_2 is ubiquitous, carbonates are formed (Fig. 20); calcium carbonate encrustations are characteristic of sulfureta in which calcium sulphate is the major sulphate source (e.g. most sulphur springs). Wadi Natrun in Egypt is a natural soda (Na_2CO_3) deposit which was studied by Jannasch (1957) and by Abd-el-Malek & Rizk (1963). Jannasch (1957) mentioned its biblical association with Moses and described its population of coloured phototrophic sulphur bacteria. It appears to be a sulfuretum bearing some resemblance to the Libyan sulphur lakes described earlier in this section, but with sodium sulphate as the primary sulphate source. Abd-el-Malek & Rizk (1963) related the production of soda to the population of *Desulfovibrio*, thus providing the best-documented account of the biogenesis of soda. Calcite deposits which often surround sulphur springs may well arise in a similiar manner; Römer & Schwartz (1965) suggested other alkaline earth carbonates might have a comparable origin.

Food spoilage

Halophilic sulphate-reducing bacteria may contaminate pickled, stoppered foods (e.g. olives in olive brine, Levin *et al.*, 1959). The spores of *Desulfotomaculum nigrificans* are very resistant to heat and these organisms cause 'sulphur stinker' spoilage of canned vegetables such as corn

Food spoilage   145

and peas (they were first discovered in this context from corn as *Clostridium nigrificans*). Sulphur stinker spoilage is characterized by a greying or, in a bad case, blackening of the food with the accompanying 'bad eggs' smell of H_2S; it is now fairly rare. Desulfotomacula can also spoil molasses, which is heated industrially to make it more mobile for handling, and have been reported as contaminants of infant food preparations (Donelly & Busta, 1980) and canned mushrooms (Lin & Lin, 1970). Postgate (1982*a*) mentioned an instance of contamination of a medicinal Al_2O_3 sol by sulphate reducers, which was controlled by adding hydrogen peroxide. In general, the control of food spoilage by sulphate-reducing bacteria is a matter of common sense: either sterilization should be complete or anaerobic conditions should be avoided. The less exotic food-spoilage organisms present the more serious problems to the food industry.

Oil technology

Sulphate-reducing bacteria appear to be indigenous inhabitants of the waters of oil-bearing shales and strata, and their role in oil technology has been the subject of many special reports, a symposium (Anderson, 1957) and features in at least two books (Beerstecher, 1954; Davis, 1967). Their activities in this context are numerous. They corrode pumping machinery, storage tanks and other installations; they generate H_2 which contaminates the natural gas associated with oil deposits; they grow in injection waters (below) and may clog the system; and they may have more fundamental roles in the formation, release and transformation of oil hydrocarbons. The problems they generate have been familiar to oil technologists for most of this century; they have arisen in new forms with the exploitation of off-shore oil reserves over the last two decades (e.g. Wilkinson, 1982). In the context of this monograph I can only indicate briefly the nature of their activities; for details the technological literature should be consulted.

Formation of oil hydrocarbons

Jankowski & ZoBell (1944) first reported oil-like products in cultures of sulphate-reducing bacteria grown in heterotrophic conditions. Sisler & ZoBell's (1951*a*) culture which produced material bearing a close chemical relationship to petroleum hydrocarbons was mentioned in Chapter 5 (p. 69). Sisler & ZoBell wrote: 'It is not to be concluded from these and similar observations that autotrophic sulphate reducers produce crude oil ...' and went on to emphasize, as did Stone & ZoBell (1952), that sulphate-reducing bacteria were probably just one of the many types of organism that produced small amounts of hydrocarbon-like material as part of their cell substance.

A question of critical importance in considering the function of sulphate-reducing bacteria in the genesis of petroleum is whether they are normal inhabitants of oil deposits. They are certainly easy to detect in most water and rock samples from oil wells; however, not only are they readily introduced by the drilling procedure, but ZoBell (1958) calculated that they could travel at between 0.6 and 12 m per year through oil-bearing sands. Thus, bacteria introduced at the original site could contaminate a wide area after a few decades. ZoBell (1958) described some ingeniously conceived experiments designed to obtain a sterile core sample, using highly alkaline drilling muds and dipping the core into molten wax as soon as it was received, as a result of which the real presence of sulphate-reducing bacteria in the core was demonstrated. But the possibility of contamination from 20- or 30-year-old neighbouring drills could not be excluded. Dostalek & Kvet (1964) demonstrated that the salt tolerance of bacteria occurring naturally in seam waters is a fairly stable property, and raised the possibility that indigenous bacteria might be distinguished from those introduced by contamination by their degree of adaptation to the salinity of the ambient water. Kutznetsova & Pantskhava (1962) and Kutznetsova, Li & Tiforova (1963) used a similar principle, deducing that the bulk of the bacteria in certain oil well waters was not indigenous because the samples required diluting before they would grow. Logic of this kind should be accepted with caution because imperfect adjustment to an extreme environment may not necessarily imply that the organism is an intruder; slow or feeble growth, which in the laboratory may appear as a sign of poor adaptation, may have relative advantages in nature. In my experience, most samples from very saline environments (e.g. salt pans, the Dead Sea) have given more rapid growth in media that were somewhat more dilute than their natural environment.

Release of oil hydrocarbons from oil-bearing shale

ZoBell (1947a) reported that, when his cultures grew in the presence of oil–sand mixtures, the oil was regularly released from the sand, though it remained adsorbed in controls in which growth did not occur. The oil was not utilized, so he concluded that the bacteria facilitated the release of adsorbed oil, and pointed out that, in nature, this process might be important in the genesis of natural reservoirs of oil. He suggested several mecha-

Oil technology 147

nisms by which this might occur (see also ZoBell, 1947b, 1950):

1. Direct action of the bacteria on the mineral: utilization of sulphate or carbonate might promote the release of oil from mineral sulphates or carbonates, though such a mechanism could not account for the release of oil adsorbed on sand.

2. Expulsion of oil by gas production: gas formed by bacteria could move oil either by blowing the oil out of a pore or by dissolving in the oil and so altering its viscosity.

3. Physical displacement of adsorbed oil droplets: sulphate-reducing bacteria and many other bacteria tend to adhere to surfaces; thus they might physically displace adsorbed oil.

4. Production of detergents, perhaps by oxidation of the oil itself.

5. Reduction of the viscosity of oil by oxidation.

ZoBell was able to demonstrate most of these processes and, despite criticism, they were essentially substantiated by the work of La Rivière (1955a, b) and Spurny & Dostalek (1958); however Updegraff & Wren (1954) flatly contradicted these findings. Doubt remains; bacteria other than sulphate-reducers produce surface-active substances (La Rivière, 1955b) but the evidence that some strains of *Desulfovibrio* do so too is reasonably good. Dostalek & Spurny (1958) and Dostalek (1961) were encouraged by these findings to inject nutrients (e.g. distillery water, molasses) into spent oil bores to stimulate the activity of sulphate-reducing bacteria, and reported transient increases in yield.

Modification of oil hydrocarbons

Bacteria might alter hydrocarbons either by reducing unsaturated molecules to saturated molecules or by oxidizing both types. Rosenfeld (1948) and ZoBell (1950) discussed the hydrogenation process – analogous to the familiar reduction of fumarate to succinate – but there is in fact no evidence for hydrogenation of a true hydrocarbon by these bacteria.

Oxidation of hydrocarbons at the expense of sulphate reduction has been discussed, and indeed reported, numerous times. Though there is little doubt that, in nature, apparently anaerobic modification of petroleum hydrocarbons does occur, the evidence for the involvement of sulphate-reducing bacteria suffers from the defect that the purity of the cultures used is sometimes questionable as is the purity of the specimens of hydrocarbon. Beerstecher (1954) also expressed doubts whether sulphate-reducing bacteria oxidize hydrocarbons. Wake *et al.* (1977) excluded most hydrocarbons as substrates on thermodynamic grounds but

148   Economic activities

predicted that lower alkynes would be thermodynamically feasible substrates. However, they did not obtain oxidation of acetylene either by pure strains or by enrichment cultures of sulphate-reducing bacteria, a negative result which can now be understood in the light of the probable inhibitory effect of acetylene on hydrogenase (p. 95).

In the oxidation of hydrocarbons by aerobes a phenomenon called 'co-oxidation' may occur (see Foster, 1962) in which methane organisms utilize propane, for example, only if methane is being utilized simultaneously; oxidation of propane alone will not support growth. Co-oxidations of hydrocarbons might account for the divergence of opinion on whether sulphate-reducing bacteria oxidize hydrocarbons; alternatively, they may be mixotrophic substrates (p. 45), i.e. energy sources but not growth substrates. At present most authorities take the view that hydrocarbons are not oxidized by sulphate-reducing bacteria, but the question is not entirely settled (see Davis & Yarbrough, 1966).

Contamination of injection waters

Much oil can be obtained from nominally exhausted wells by injecting water into the formation to displace residual oil, a process called 'secondary recovery'. Such procedures usually stimulate growth of indigenous microbes, including sulphate-reducing bacteria, which tend to clog the formation and diminish the economic yield. Herbert (1976) mentioned that North Sea oil, when in full production, may require the injection of 36×10^8 litres of sea water per day which must be filtered free of particles, de-aerated (since aerated sea water is highly corrosive) and compressed to between 20 250 and 121 550 kPa. So treated, the sea water will become a favourable environment for barotolerant marine *Desulfovibrio* species and the injection water could become both corrosive and polluted. Control of this situation is difficult and, in a marine context, next to impossible. The varieties of quaternary and other inhibitors (see Appendix 2) that have been considered were largely developed in response to the demand for a bactericide to prevent bacterial plugging of injection systems; for the latest means of control, the specialized (and regrettably often secretive) literature should be consulted.

Spoilage of stored oil products

Bulk supplies of petroleum and kerosene are normally stored in tanks which have, sometimes fortuitously, water beneath the hydrocarbon phase. Often this is sea water, since many primary storage tanks are

Oil technology 149

located beside the sea so they may be filled directly from oil tankers; in that situation a pulse of sea water usually precedes oil during the filling operation. Reduction of sulphate in the water bottom leads to sulphide contamination of the oil layer and, in normal circumstances, a sol of sulphur is formed by autoxidation. High grade petroleum can be seriously polluted by this process and, in the era of piston-engined aircraft, serious economic loss arose from down-grading of aviation fuel in sub-tropical areas and at least one instance of strategic embarrassment to military aircraft is on record in 1956 (see Postgate, 1960a). Modern jet aircraft are more tolerant in their fuel requirements, but care in the management of fuels so stored is still necessary. Chromate (Postgate, 1960a) or borate (Crombie, Moodie & Thomas, 1980) in the water layer have been recommended as prophylactics, but a rapid turnover of fuel is probably most effective because it prevents anaerobic conditions from developing. The carbon source for sulphate reduction is probably not the hydrocarbon itself but rather oxidation products formed by aerobic hydrocarbon-oxidizing bacteria which are always plentiful in the aqueous substratum. They have also been found in the watery layer of aircraft fuel tanks, but I know of no record of sulphate-reducers surviving there.

Iron sulphide is formed in such environments, as a characteristic black deposit. It may be so fine as to be pyrophoric and present a hazard: Watts (1936, 1938) reported cases of explosions due to ignition of pyrophoric FeS during cleaning-out of petroleum storage tanks.

'Cutting emulsions' are suspensions of oil in water which are used to cool and lubricate tools in industrial metal-working operations. They can become infected with microbes, including pathogens, which cause problems including allergy, spread of disease, discolouration and breakdown of the emulsions (Hill, 1975). In addition to mineral oil, such emulsions contain fatty oils, emulsifiers, fatty acids, rosin, naphthenic acids, alcohols or high-boiling ethers, rust inhibitors and 'sulphurized fatty oils'. Although marketed sterile, such components are good substrates for aerobic bacteria (Sabina & Pivnick, 1956). In addition, careless handling may make bacterial contaminations worse: Bennett (1956a) was moved to recommend that 'workers should not be allowed to spit, urinate or throw portions of their lunches into the emulsion'. Spoilage involving sulphate-reducing bacteria takes place typically during stagnant periods, over week-ends for example, and is characterized by breakdown of the emulsion, blackening and a smell of H_2S (Liberthson, 1945; Bennett, 1956b; see also Guynes & Bennett, 1959; Isenberg & Bennett, 1959). Aeration of the emulsion prolongs its life.

Contamination of gas

Town gas, whether originating as natural gas or 'producer gas', is often stored in gas holders over water. Growth of sulphate-reducing bacteria in the water layer causes a problem which has faced the gas industry for many decades (Senez, Geoffrey & Pichinoty, 1956; Pankhurst, 1968a, b): the gas becomes contaminated with H_2S, is offensive and, more seriously, generates damaging amounts of sulphur dioxide when burned. Corrosion of piping, valves and jets may also occur. In gas holders, their activity can be controlled economically with microbicides because the water volumes are relatively small, but the problem has become less treatable with the increased use of natural gas, which is stored in subterranean aquifers (Pankhurst, 1968b).

Paper technology

Blackening of paper pulp by sulphate-reducing bacteria was recognized several decades ago (Beckwith & Moser, 1932). They can spoil pulp, so that the paper then contains grey flecks, and render paper unsuitable for photography. The sulphite-containing waste liquor of the paper industry is, of course, a marvellous culture medium for sulphate-reducing bacteria and can cause a serious disposal problem. The disposal of paper pulp wastes into the Androscoggin River, USA, required the simultaneous additions of 4.5 tonnes of nitrate per day to prevent sulphide pollution (Lawrance, 1948, 1950). A procedure exploiting sulphate reduction to treat such effluents was mentioned earlier under 'Purification of wastes' (see Bannink & Muller, 1952). The involvement of sulphate-reducing bacteria in the paper industry was reviewed by Starkey (1961) and Russell (1961); the latter reviewer described a complex involvement: logs for paper manufacture which are stored in sea water can become saturated with sea salts and infected with halotolerant *Desulfovibrio*. The sulphide formed by these bacteria converts some of the lignin to thiolignin. When the wood is pulped, this thiolignin can react with anti-fungal mercurials so that the pulp rots unexpectedly. The uncovering of the sequence was a fascinating piece of microbiological detective work.

Animal nutrition

Sulphate-reducing bacteria are normal, if minor, components of the microflora of the rumen (Lewis, 1954; Guttierez, 1953) although there have been uncertainties about their characterization (compare Huisingh, McNiell & Matrone, 1974, with Howard & Hungate, 1976). *Desulfotomaculum acetoxidans* is found in the intestinal contents of higher animals (p. 20). *Desulfovibrio* can often be found in rumen fluid and *Desulfotomaculum ruminis* was first isolated from the rumen of a hay-fed sheep (Coleman, 1960); it is a diazotroph (Postgate, 1970a) but it seems unlikely that it, or any other rumen diazotroph, contributes significantly to the nitrogen status of its host (Hobson *et al.*, 1973). On the other hand, there is some evidence (Anderson, 1956) that sulphate-reducing bacteria contribute to the sulphur nutrition of ruminants. They may also play a part in the nutrition of insects such as termites (Haines, Henry & Block, 1960; see also Postgate, 1960a). They can play an important part in the recently-discovered eco-systems based on the sulfuretum (see p. 118, also Postgate, 1982a), but none has yet been shown to be of serious economic importance, at least in the present era of this planet's history (cf. Chapter 6).

Pathogenicity

The sulphate-reducing bacteria are normally non-pathogenic to man and may be found in fresh human faeces (K. R. Butlin, personal communication). Sefer & Calinescu (1969) found *Desulfovibrio* among oral streptococci in seven specimens of human dental caries but their role in the infection was not clear. I have heard of only one appearance of these bacteria in a systemic clinical condition: in 1976 Dr Richard Porschen of a hospital in Long Beach, California, USA, sent me a culture of a typical desulfoviridin-positive *Desulfovibrio* isolated from the blood of a patient under treatment for a variety of symptoms. The *Desulfovibrio* bacteriaemia was associated with fever from which the patient recovered spontaneously within 24 hours.

Miscellaneous economic activities

Blackening of metal-based paints and pigments, of metalwork and materials containing traces of metal (e.g. paper pulp) has already been discussed in various contexts. I am sure I have not covered all possible economic activities in which such problems arise: one not previously mentioned is the blackening of leather during tanning, for which sulphate-reducing bacteria have been held responsible (Krassowski *et al.*, 1966). No doubt many other instances of spoilage, corrosion and nuisance could have been cited, but it ought now to be clear that the economic activities of this group of microbes are exceptionally widespread and diverse. The possibilities of exploiting these bacteria for biotechnological processes have interested many scientists – the microbiological production of sulphur or sulphide and the rejuvenation of spent oil wells are two clear examples which have reached advanced experimental stages but which have not (so far) proved truly economic.

An area of biotechnological interest which has emerged since the world energy shortage became widely recognized is the biological production of H_2 as an energy source. Photolysis of water to H_2 ($+O_2$) on a mass-production scale, making use of a chloroplast (or chloroplast-like) system coupled to a ferredoxin and hydrogenase, is attractive as a method of storing solar energy. *D. vulgaris* possesses one of the most active hydrogenases known (p. 72), and this could well be the enzyme of choice for any viable biotechnological process because it can be extracted in an O_2-tolerant form (van der Westen, Mayhew & Veeger, 1980). Not all desulfovibrios have such active hydrogenases. Yet another area in which these bacteria might be biotechnologically exploitable is in biochemical fuel cells. Organisms which generate a redox gradient are particularly suitable for such use; it will be obvious that the sulphate-reducing bacteria lower the redox potential so a culture of these bacteria makes a plausible half-cell for an oxido-reduction battery. The use of sulphate-reducing bacteria in redox biochemical fuel cells was discussed by Sisler (1961), Sisler & Senftle (1963) and Lewis (1966); Sisler, Senftle & Skinner (1977) combined such a redox cell with a pH-gradient cell involving thiobacilli in a process for the electrobiochemical neutralization of acid mine water.

9

Epilogue

For a group of microbes so manifestly ill-adapted to the normal aerobic terrestrial environment, the sulphate-reducing bacteria show an amazing capacity for survival and an almost anthropomorphic determination to recover their putative ancient role in the biosphere whenever circumstances permit. They exert a strange fascination on any microbiologist who has handled them, largely because of the glimpses they give of a biochemistry as different from the main stream of terrestrial life as can reasonably be expected – if one accepts the view that terrestrial living things have a common origin. Indeed, their biochemical eccentricity has made at least a few scientists see them as models for conceivable extraterrestrial biota (e.g. Postgate, 1970b; Clark, 1979, 1981). The surface regolith of the planet Mars is apparently rich in sulphate sulphur (Clark & Baird, 1979; Clark & van Hart, 1981).

Their isolation and cultivation is no longer the problem it was in the 1930s, but there are still times when the microbiologist has to admit he is baffled: good black colonies of an isolate will doggedly refuse to yield liquid cultures, or a familiar strain, *Desulfovibrio gigas* or *Desulfotomaculum orientis* perhaps, will capriciously give poor growth or, in the latter case, unexpected mass sporulation. Often the trouble resolves itself for reasons which elude the experimenter; on at least two occasions I am aware of Divine intervention being sought to ensure growth of *D. orientis*! But on the whole the sulphate-reducing bacteria have been 'tamed'. Yet despite the impressive advances in our knowledge of them during the last two decades, even elementary questions remain unanswered. Do they really attack hydrocarbons? How far does their carbon metabolism resemble that of aerobes? Why does the carbon metabolism of so many species stop at the acetate level of oxidation, when oxidation of acetate coupled to sulphate reduction is thermodynamically exergonic? Why does acetate utilization correlate with slow growth? Why does *Desulfovibrio* combine cytochromes, ferredoxins and a positive menagerie of other electron transport substances in its metabolism? Are the sulphate-reducing bacteria related to each other in any phylogenetic sense? Why, above all, are they anaerobes at all, even exceptionally exacting ones, when they have all the necessary equipment to perform a conventional oxygen-based metabolism? A lot has yet to be learned about this unique group of microbes.

Appendix 1

Characters of certain strains of sulphate-reducing bacteria held in National collections of bacteria

The British National Collection of Industrial and Marine Bacteria (NCIMB) and the Deutsche Sammlung von Mikroorganismen (DSM) between them hold representatives of the majority of sulphate-reducing bacteria discussed in this monograph, as well as some strains with special characters which were not mentioned. This appendix has been prepared to supplement the rather limited information available in their catalogues.

I. National Collection of Industrial and Marine Bacteria.
In November 1959 the NCIMB assumed responsibility for a substantial collection of sulphate-reducing bacteria from the Teddington laboratory of the late K. R. Butlin. Freeze-dried cultures of these strains may be had, for a fee, on application to the Curator, National Collection of Industrial and Marine Bacteria Ltd, Torry Research Station, 135 Abbey Road, Aberdeen, AB9 8DG, Scotland (telephone: Aberdeen (0224) 877071). The donors of the strains and references thereto are published in the third edition of NCIMB's *Catalogue of strains* (1975, HM Stationery Office) and will not be repeated here. For ease of comparison with the NCIMB catalogue, the strains are listed in order of their NCIMB numbers within genera. Strain 8322, listed in the 1975 catalogue, was no longer available at the time of writing. All source locations are in Britain except where stated; salt concentrations (where quoted) and temperatures are those routinely used for culture. References to most of the strains here are also to be found in Postgate & Campbell (1966) and Campbell & Postgate 1965). I thank Mr Angus MacKenzie of the NCIMB for checking this list and assisting with the commentary.

Appendix 1

NCIMB No.	Strain name	Commentary
Genus *Desulfotomaculum*		
Desulfotomaculum nigrificans		
8351	Teddington garden	Isolated in late 1930s from garden soil outside the laboratory at Teddington. 55 °C. Motile. Typical morphology. Hydrogenase not found.
8353	Staines G	From soil sample near Staines, Middlesex, in 1948. 55 °C. Typical morphology.
8354	Aberdeen	From soil near Aberdeen, 1948. 50 °C. Hydrogenase not found.
8355	Holland DT	Probably ditch water in Holland, 1937. 55 °C. Hydrogenase not found.
8395	Delft 74	55 °C. Starkey's strain '74T'. Hydrogenase present. Sporulates more readily than most desulfotomacula. Suggested neotype strain.
8706	None	Isolated from molasses in the 1950s as *Clostridium nigrificans*. 55 °C.
8788	None	As 8706. 55 °C. Poor motility.
Desulfotomaculum orientis		
8382	Singapore I	From soil from Rangoon Road pumping station near rising main, Sungei Whampoa, Singapore, 1955. 30 °C. Holotype strain. Curved morphology, slight motility.
8445	Singapore II	Mud from Ulu Pandan, 6 km from Sungei Whampoa, Singapore, 1958. 30 °C.
Desulfotomaculum ruminis[1]		
8452	DL	Holotype strain, from hay-fed sheep. 37°C.
10149	Coleman 42	A co-type of 8452. 37 °C.

[1]See Campbell & Postgate (1969) concerning a confusion of holotype and neotype strains. The listing here is correct.

Characters of sulphate-reducing bacteria 157

NCIMB No.	Strain name	Commentary
Genus *Desulfovibrio*		
Desulfovibrio africanus		
8397	Walvis Bay	Water sample from Walvis Bay, South Africa, after an 'eruption' of H_2S. 30 °C, 2.5% NaCl. Slender vibrio morphology. Deposited as '*D. aestuarii*'.
10470		From water sample, Pematty Lagoon, South Australia, 1970. 30 °C. Hydrogenase present. Sigmoid morphology.
8401	Benghazi 1	Holotype strain. From well water from Benghazi, Libya, 1950–51. 31 °C. Slender sigmoid vibrios.
Desulfovibrio desulfuricans subsp. *aestuarii*		
9335	Sylt 3	Marine origin from Germany. Holotype of subspecies *aestuarii*. Obligate halophile, 1961.
Desulfovibrio desulfuricans subsp. *desulfuricans*		
8307	Essex 6	Holotype strain. From sand and tar mixture round a corroded gas main in clay; south Essex. 30 °C.
8310	Norway 4	From water from Oslo Harbour, Norway, 1949. Straight cells which lack desulfoviridin. Otherwise normal.
8312	Teddington R	River mud at Teddington Lock by H. J. Bunker, 1939. One of the few non-motile strains of *Desulfovibrio* (but see p. 21). 30 °C.
8313	New Jersey Sol C-1	Interior of corroding cast-iron heat exchanger, 1940s. 30 °C.
8318	El Agheila Z	The same isolate as 8380 acclimatized to malate media. 30 °C.
8326	California 27.137.5	Top soil of an oil field, Long Beach, California, USA. Purified 1948–50. 30 °C. Requires NaCl.
8338	Cuba HC 29.130.4	From estuary well. Purified 1949. 30 °C. Halophile.

NCIMB No.	Strain name	Commentary
8339	California HC 29.137.11	From oil well brine. 30 °C, 2.5% NaCl.
8363	Canet 40	From Etang de Canet, Perpignan, France. 30 °C, 2.5% NaCl.
8372	Holland C-6 or America	This is ATCC strain 7757, isolated by R. L. Starkey from canal mud at Delft, Holland, c. 1938. 30 °C.
8380	El Agheila Z	Sulphurous mud from Ain-ez-Zauia lake, near El Agheila, Libya, 1950–51. 30 °C, 2.5% NaCl. See also 8318.
8387	Berre-eau	Water sample from Etang de Berre, Marseille, France. 30 °C. Good nitrogen fixer. Straight cells.
8388	Berre-sol	Soil sample from Etang de Berre, Marseille, France. 30 °C. Well-established nitrogen fixer. Straight cells.
8393	Canet 41	As 8363.
8400	Hossegor	Tidal marine lake from Hossegor, southwest France. 30 °C, 2.5% NaCl.
8449	Zurich	30 °C.
9467	'Vibrio cholinicus'	Isolated in USA as choline-utilizing '*V. cholinicus*'. Re-purified 1972. 30 °C.
9492	Aberdovey	Mud from River Dovey, Wales, near Dovey railway junction, 1963. 30 °C.

Desulfovibrio gigas

9332		Holotype strain. Etang de Berre, Marseille, France. 30 °C. Unusually large, spirilloid morphology.

Desulfovibrio salexigens

8308	El Agheila C	From dried mud samples from Ain-el-Braghi, a sulphur lake near El Agheila, Libya. Received and isolated 1953. 30 °C, 2.5% NaCl.
8315	New Jersey SW8	Probably marine sediment or steel corroding in sea. 30 °C. 2.5% NaCl.

Characters of sulphate-reducing bacteria 159

NCIMB No.	Strain name	Commentary
8329	Australia	Sample of water from Lake Eyre, Australia, 1956. 30 °C, 2.5% NaCl.
8364	California 43 : 63	From mud sample from Sorrento Slough, near Delmar, California, USA. 30 °C, 2.5% NaCl. Supplied by Professor ZoBell as a fumarate-reducing type.
8365	Louisiana 43 : 11	From mud cove on Timbalier Island, Louisiana, USA. 30 °C, 2.5% NaCl. Supplied by Professor ZoBell as a carbonate reducer.
8402	El Agheila 2	Sulphur sample from Ain-ez-Zauia, near El Agheila, Libya, 1950–51. 30 °C, 2.5% NaCl. Forms mucin in culture.
8403	British Guiana	Holotype strain from 'sling mud' from British Guiana, 1951–52. 30 °C, 2.5% NaCl.

Desulfovobrio vulgaris subsp. *oxamicus*

9442	Monticello 2	Mud sample from polluted stream at Monticello, near Urbana, Illinois, USA. Utilizes oxamate for mixotrophic growth. 30 °C.

Desulfovibrio vulgaris subsp. *vulgaris* (several strains were deposited as *D. desulfuricans*)

8302	Teddington M	River mud from Teddington Lock, 1938. 30 °C.
8303	Hildenborough	Holotype strain. From clay near Hildenborough, Kent, 1946. 30 °C.
8306	Wandle	From river water at Wandsworth, London, 1944–46. 30 °C.
8311	Holland D6	Ditch mud in Holland by R. L. Starkey, 1938. One of the few non-motile strains of *Desulfovibrio*. 30 °C.
8386	Marseille-gaz 54	From gas holder water, Marseille, France, 1958. 30 °C. Claimed to be able to reduce nitrate.

NCIMB No.	Strain name	Commentary
8399	Venezuela	Oily, H_2S-bearing mud from Maracaibo Lake, Venezuela. 30 °C, 2.5% NaCl.
8446	Llanelly	Soil by corroded gas main, Llanelly, 1958. 30 °C.
8456	Denmark	
8457	Woolwich	Slime covering corrosion test specimens in River Thames at Woolwich, London, 1960. 30 °C.
11727	Marburg	Anaerobic sewage digester, Marburg, Germany, by R. Thauer and colleagues. Non-motile. 30 °C.
11564	F2	Filtered and chlorinated sea water from an oil-production platform in Forties Field in North Sea. 1979. 30 °C.
11565	S6	Untreated sea water from an oil-production platform in Forties Field in North Sea. 1979. 30 °C.
11779	Groningen	A non-motile strain of remarkable semi-lunar morphology isolated by H. Veldkamp from fresh water polluted by a starch factory. 30 °C.

Desulfovibrio spp.[1]

8301	Holland SH-1	From Dutch clay soil of uncertain origin. Impure culture from R. L. Starkey in 1937 purified by M. E. Adams, 1949. 30 °C.
8306	Brockhurst Hill	From a tubercle in a water main at Brockhurst Hill, 1948. Said by M. E. Adams to grow best at pH 8.5. 30 °C.
8314	New Jersey SW7	Uncertain marine origin. Rather long cells. 30 °C, 2.5% NaCl.
8320	New Jersey RM4	From river mud about 1945. 30 °C.
8338	Cuba HC 29.130.4	30 °C, 2.5% NaCl.

[1] Not identified at species level. All except 10455 listed in NCIMB catalogue of 1975 as *D. desulfuricans*.

NCIMB No.	Strain name	Commentary
8374	Clyde II	Black oily mud from River Cart at its confluence with R. Clyde at Renfrew near Glasgow. 30 °C, 2.5% NaCl.
10455		Deposited as *Desulfovibrio fluorescens*.

II. Deutsche Sammlung von Mikroorganismen.

The German collection of micro-organisms holds relatively few sulphate-reducing bacteria but their collection includes representatives of most of the 'new' genera discovered by Professor Pfennig and Dr Widdel as well as the only known second isolate of *Desulfovibrio gigas*. Cultures may be obtained, for a fee, from the Deutsche Sammlung von Mikroorganismen, D-34 Göttingen, Griesebachstrasse 8, West Germany. I thank Dr F. Widdel, who prepared the greater part of this list.

DSM No.	Strain name	Commentary
	Genus *Desulfotomaculum*	
Desulfotomaculum acetoxidans		
771	Göttingen	Monotype strain, from piggery waste. 36 °C. Sporulation with concomitant formation of gas vesicles.
	Genus *Desulfovibrio*	
Desulfovibrio gigas		
496		Schoberth's (1973) strain isolated from Göttingen sewage.
Desulfovibrio baarsii		
2075	Konstanz	Monotype strain, from a ditch in Konstanz. Grown at 38 °C.
Desulfovibrio sapovorans		
2055	Lindhorst	Monotype strain, from a village ditch near Hannover. Grown at 34 °C.
	Genus *Desulfobacter*	
Desulfobacter postgatei		
2034	Dangast	Monotype strain, from a brackish water ditch near Jadebusen, North Sea. Grown at 32 °C with 1% NaCl and 0.05% $MgCl_2$.

DSM No.	Strain name	Commentary
	Genus *Desulfobulbus*	
Desulfobulbus propionicus		
2032	Lindhorst	Monotype strain, from a village ditch near Hannover. Grown at 39 °C.
	Genus *Desulfococcus*	
Desulfococcus multivorans		
2059	Göttingen	Monotype strain, from sludge digester, Göttingen. Grown at 35 °C, halotolerant.
	Genus *Desulfonema*	
Desulfonema limicola		
2076	Jadebusen	Monotype strain, from mud flat of the Jadebusen, North Sea. Gliding multicellular filaments. Grown at 30 °C with 1.5% NaCl and 0.1% $MgCl_2$.
Desulfonema magnum		
2077	Montpellier	Monotype strain, from sediment of a Mediterranean lagoon, Montpellier. Gliding multicellular filaments. Grown at 32 °C with 2% NaCl, 0.25% $MgCl_2$ and $CaCl_2$. Long term preservation in liquid nitrogen difficult.
	Genus *Desulfosarcina*	
Desulfosarcina variabilis		
2060	–	Monotype strain, from sediment of a Mediterranean lagoon, Montpellier. Cell packets or single oval cells. Grown at 33 °C with 1.5% NaCl and 0.1% $MgCl_2$. Sensitive to light.

Appendix 2

Selected list of inhibitors of sulphate-reducing bacteria

Largely condensed from a more extensive compilation by Saleh *et al.* (1964) which should be consulted for documentation. $Dv = Desulfovibrio$ *vulgaris*; $Dd = Desulfovibrio$ *desulfuricans*; $Dg = Desulfovibrio$ *gigas*; $Dn = Desulfotomaculum$ *nigrificans*; $Do = Desulfotomaculum$ *orientis*. All test media are based on lactate, sulphate, the inhibitor and a reductant, except where mentioned.

Appendix 2

Substance	Minimum inhibitory concentration (MIC) (μg/ml)	Organism	Temperature (°C)	Commentary
Quaternaries				
American Petroleum Institute quaternary	> 75	Impure	37	Arquads are commercial quaternaries of uncertain structure produced by Armour Inc.
'Arquad 16' 50%	0.25	Dn	55	
'Arquad 16' 50%	10	Dv	30	
'Arquad 2C' 50%	5	Dn	55	
'Arquad 2C' 50%	50	Dv	30	
'Arquad S-2C' 50%	0.2	Dn	55	
'Arquad S-2C' 50%	10	Dv	30	
Cetyltrimethylammonium bromide	1	Dn	55	Resistance augmented by NaCl (Costello et al., 1970)
Cetyltrimethylammonium bromide	10	Dv	30	
Dimethylbenzyllaurylammonium chloride	5	Dd	37	Tested as bactericide over 15 min
Stearyldipolyglycolbenzylammonium chloride	5	Dd	37	Tested as bactericide over 15 min
Cetylpyridinium bromide	10	Dd	30	No reducing agent
Antibiotics				
Chlortetracycline	100	Dn	55	
Chlortetracycline	100	Dv	30	
Chloramphenicol	5	Dn	55	
Chloramphenicol	50	Dv	30	

List of inhibitors of sulphate-reducing bacteria 165

Dihydrostreptomycin	1	Impure	22–25	No reducing agent
Erythromycin	1	Impure	22–25	No reducing agent
Neomycin	140 units/ml	Dv	30	No reducing agent
Penicillin	$> 10^4$ units/ml	Dn	55	No reducing agent
Penicillin	$> 10^3$ units/ml	Dv	30	No reducing agent
Polymyxin B	10	Dn	55	
Polymyxin B	100	Dv	30	
Streptomycin	500	Dn	55	
Streptomycin	$> 10^3$	Dv	30	
Tetracycline	20	Impure	22–25	No reducing agent
Dyes				
Acridinium dye 914	10	Dv	30	No reducing agent, peptone medium
Proflavine	100	Dn	55	
	10	Dv	30	
Benzyl viologen	50	Dn	55	
	50	Dv	30	
Crystal violet	10	Impure	37	
Methylene blue	$> 10^3$	Dv	30	
	10^3	Dn	55	
Mercurials				
o-chloromercuriphenol	500	Dd	30	
p-chloromercuritoluene	$> 10^3$	Dd	30	
ethyl mercuric acetate	10^3	Dd	30	
Ethyl mercurithiosalicylate	250	Dd	30	
Mercuric chloride	10^3	Dd	30	
Mercury naphthenate	10^3	Dd	30	
Phenyl mercuric acetate	250	Dd	30	Eleven other phenyl mercuric salts were rather less active

Substance	Minimum inhibitory concentration (MIC) (µg/ml)	Organism	Temperature (°C)	Commentary
Phenyl mercuri-8-hydroxy quinolate	100	*Dd*	30	The most active phenyl mercuri-salt
'Menaphthan'	100	*Dd*	30	4-nitro-5-hydroxymercuri-*o*-cresol anhydride
Metal ions				
Copper sulphate	5–30	*Dn*	55	
Copper sulphate	20	*Dd*	30	2.5% NaCl in medium
Copper sulphate	5–50	*Dv*	30	
Copper sulphate	30	*Do*	30	
Copper 8-hydroxyquinolate	250	*Dn*	55	
Copper 8-hydroxyquinolate	250	*Dv*	30	
Zinc sulphate	10^4	*Dd*	30	No reducing agent
Nitro compounds				
2-bromo-2-nitropropyl acetate	100	*Dd*	30	
m-dinitrobenzene	16	Impure	—	Sewage samples
Nitrobenzene	16	Impure	—	Sewage samples
2-nitro-1-butanol	50	*Dd*	30	
2-nitro-2-ethyl-1, 3-propanediol	100	*Dd*	30	
'Furacin'	100	*Dn*	55	5-nitro-2-furaldehyde semicarbazone
	100	*Dv*	30	
Picric acid	8	Impure	—	Sewage samples
Trinitrotoluene	18	Impure	—	Sewage samples

List of inhibitors of sulphate-reducing bacteria

Phenolics				
Phenol	10^4	Dv	30	
6-chloro-thymol	25	Dn	55	
	100	Dv	30	
m-cresol	$> 10^3$	Dd	30	
Oxine	500	Dn	55	No reducing agent
	500	Dv	30	8-hydroxyquinoline
β-naphthol	> 100	Dn	55	
	> 100	Dv	30	
Octyl cresol	Saturated	Dd	30	No reducing agent
Pentachlorophenol	5	Dn	55	
	50	Dv	30	
2-bromo-4-phenylphenol	25	Dd	30	The most active of 62 substituted
2-chloro-4-nitrophenol	25	Dd	30	phenols recorded by Saleh et al. (1964)
Tannins	$> 10^4$	Dv	30	
Miscellaneous				
'Maphenide'	500	Dn	55	A sulphonamide drug
	500	Dv	30	
Sulphanilamide	> 10	Dv	30	No reducing agent
'Sulfathiazole'	500	Dv	30	No reducing agent
4-aminoquinaldinium decyl-acetate	1	Dn	55	
	100	Dv	30	
'Hibitane'	0.1	Dn	55	bis-p-chlorophenyldiguanido hexane diacetate
'Hibitane'	1	Dv	30	Decamethylenebis-4-amino-
Dequadin acetate	1	Dn	55	quinaldinium acetate
Dequadin acetate	100	Dv	30	
'2 : 4 D'	> 100	Dn	55	2,4-dichlorophenoxyacetic acid

Substance	Minimum inhibitory concentration (MIC) (μg/ml)	Organism	Temperature (°C)	Commentary
'2 : 4 D'	>100	Dv	30	
'Panacide'	0.5	Dn	55	2,2'-dihydroxy-5,5'-dichlorodiphenyl methane
'Panacide'	15	Dv	30	
EDTA	10^4	Dn	55	Ethylenediaminetetra-acetic acid
EDTA	>1/8	Dv	30	
Formaldehyde	60	Dd	37	
Glutaraldehyde	100	Dn	55	
	100	Dv	30	
Acetylene[a]	(20%)	Dd	30	10% was partial inhibitor
		Dg		Dm. ruminis not inhibited
Lauryl ammonium acetate	5	Dd	37	Tested as bactericide
Lauryl dimethylamine oxide	10	Dd	37	Tested as bactericide

Saleh et al. (1964) listed about 30 more miscellaneous substances of generally low inhibitory activity.
[a] Payne & Grant (1982). Acetylene in gas phase.

References

[Note: numbers preceded by T (e.g. T711–20) after a Russian literature reference refer to the pages in the English language translation of the journal.]

Abd-el-Malek, Y. & Rizk, S. G. (1963). Bacterial sulphate reduction and the development of alkalinity. III. Experiments under natural conditions in the Wadi Natrun. *J. appl. Bact.*, **26**, 20–6.
Abdollahi, H. & Nedwell, D. B. (1980). Serological characteristics within the genus *Desulfovibrio*. *Ant. van Leeuwenhoek*, **46**, 73–83.
Abram, J. W. & Nedwell, D. B. (1978a). Inhibition of methanogenesis by sulphate reducing bacteria competing for transferred hydrogen. *Arch. Microbiol.*, **117**, 89–92.
— (1978b). Hydrogen as a substrate for methanogenesis and sulphate reduction in anaerobic salt marsh sediment. *Arch. Microbiol.*, **117**, 93–8.
Ackrell, B. A., Asoto, R. N. & Mower, H. F. (1966). Multiple forms of bacterial hydrogenases. *J. Bact.*, **92**, 828–38.
Adams, M. E., Butlin, K. R., Hollands, S. J. & Postgate, J. R. (1951). The role of hydrogenase in the autotrophy of *Desulphovibrio*. *Research, Lond.*, **4**, 245.
Adams, M. E. & Postgate, J. R. (1959). A new sulphate-reducing vibrio. *J. gen. Microbiol.*, **20**, 252–7.
— (1961). On sporulation in sulphate-reducing bacteria. *J. gen. Microbiol.*, **24**, 291–4.
Adman, E. T., Sieker, L. C., Jensen, L. H., Bruschi, M. & Le Gall, J. (1977). A structural model of rubredoxin from *Desulfovibrio vulgaris* at 2 Å resolution. *J. mol. Biol.*, **112**, 113–20.
Akagi, J. M. (1967). Electron carriers for the phosphoroclastic reaction of *Desulfovibrio desulfuricans*. *J. biol. Chem.*, **242**, 2478–83.
— (1981). Dissimilatory sulphate reduction, mechanistic aspects. In *Biology of inorganic nitrogen and sulfur*, ed. Bothe, H. & Trebst, A. pp. 178–187. Springer: Berlin, Heidelberg, New York.
Akagi, J. M. & Campbell, L. L. (1962). Studies on thermophilic sulphate-reducing bacteria. III. Adenosine triphosphate-sulfurylase of *Clostridium nigrificans* and *Desulfovibrio desulfuricans*.. *J. Bact.*, **84**, 1194–1201.
Akagi, J. M. & Jackson G. (1967). Degradation of glucose by proliferating cells of *Desulfotomaculum nigrificans*. *Appl. Microbiol.*, **15**, 1427–30.
Allen, L. A. (1949). The effects of nitro compounds and some other substances on the production of hydrogen sulphide by sulphate-reducing bacteria in sewage. *Proc. Soc. Appl. Bact.*, (2), 26–38.

References

Alvarez, M. & Barton, L. (1977). Evidence for the presence of phosphoriboisomerase and ribulose 1 : 5-diphosphate carboxylase in extracts of *Desulfovibrio vulgaris*. *J. Bact.*, **131**, 133–5.

Ambler, R. P. (1968). Amino-acid sequence of cytochrome c_3 from *Desulphovibrio vulgaris* (NCIB8303). *Biochem. J.*, **109**, 47–8P.

Ambler, R. P., Bruschi, M. & Le Gall, J. (1971). Biochimie compareé des cytochromes des bactéries sulfato-réductrices. In *Recent Advances in Microbiology*, 10th Intern. Congr. Microbiol., Mexico City, pp. 25–34. Asociacón Mexicana de Microbiología.

Anderson, C. M. (1956). The metabolism of sulphur in the rumen of the sheep. *New Zealand J. Sci. Technol.*, **37**, 379–92.

Anderson, K. E. (1957). *Sulfate-reducing bacteria, their relation to the secondary recovery of oil*. New York: St. Bonaventure Univ.

Apse, J., Morwood, J. B. & Wood, S. J. (1955). Trade waste aids wet tipping control. *Muncipal J*, **63**, 371–3.

Ault, W. U. & Kulp, J. L. (1959). Isotopic geochemistry of sulphur. *Geochim. Cosmochim. Acta*, F0316, 201–35.

Baars, J. K. (1930). *Over sulfaat reductie door bacterien*. Doctoral dissertation. Meinema: Delft.

Baas-Becking, L. G. M. & Parks, G. S. (1927). Energy relations in the metabolism of autotrophic bacteria. *Physiol. Rev.*, **7**, 85–106.

Badziong, W., Ditter, B. & Thauer, R. K. (1979). Acetate and carbon dioxide assimilation by *Desulfovibrio vulgaris* (Marburg) growing in hydrogen and sulfate as sole energy source. *Arch. Microbiol.*, **123**, 301–5.

Badziong, W. & Thauer, R. K. (1978). Growth yields and growth rates of *Desulfovibrio vulgaris* (Marburg) growing on hydrogen plus sulfate and hydrogen plus thiosulfate as the sole energy sources. *Arch. Microbiol.*, **117**, 209–14.

— (1980). Vectorial electron transport in *Desulfovibrio vulgaris* (Marburg) growing in hydrogen plus sulfate as sole energy source. *Arch Microbiol.*, **125**, 167–74.

Badziong, W., Thauer, R. K. & Zeikus, J. G. (1978). Isolation and characterization of *Desulfovibrio* growing on hydrogen plus sulfate as the sole energy source. *Arch. Microbiol.*, **116**, 41–9.

Baker, K. (1968). Low cost continuous culture apparatus. *Lab. Pract.*, **17** 817–24.

Baliga, B. S., Vartak, H. G. & Jagannathan, V. (1961). Purification and properties of sulfurylase from *Desulfovibrio desulfuricans*. *J. scient. ind. Res. (India)*, **20C**, 33–40.

Banat, I. M., Lindström, E. B., Nedwell, D. B. & Balba, M. T. (1981). Evidence for the coexistence of two distinct functional groups of sulfate-reducing bacteria in saltmarsh sediment. *Appl. env. Microbiol.*, **42**, 985–92.

Bando, S., Matsuura, Y., Tanaka, N., Yasuoka, N., Kakudo, M., Yagi, T. & Inokuchi, H. (1979). Crystallographic data for cytochrome c_3 from two strains of *Desulfovibrio vulgaris*, Miyazaki. *J. Biochem.* **86**, 269–72.

Bannink, H. F. & Muller, F. M. (1952). Utilization of waste liquors from the digestion of straw with monosulfite. *Ant. van Leeuwenhoek*, **18**, 45–54.

Barghoorn, E. S. & Nichols, R. L. (1961). Sulfate-reducing bacteria and pyritic sediments in Antarctica. *Science*, **134**, 190.

Barnes, R. O. & Goldberg, E. D. (1976). Methane production and consumption in anoxic marine sediments. *Geol.*, **4**, 297–300.

Baross, J. A. & Deming, J. W. (1983). Growth of 'black smoker' bacteria at temperatures of at least 250°C. *Nature, Lond.*, **303**, 423–6.

Barta, J. (1964). Decontamination of industrial effluents by means of anaerobic continuous action of reducing sulphur bacteria. In *Continuous culture of micro-organisms*, ed. I. Malek, K. Beran & J. Hospodka, pp. 325–27. London: Academic Press.

Barton, L. L., Le Gall, J. & Peck, H. D. (1970). Phosphorylation coupled to oxidation of hydrogen with fumarate in extracts of the sulfate-reducing bacterium *Desulfovibrio gigas*. *Biochem. biophys. Res. Commun.*, **41**, 1036–42.

Barton, L. L. & Peck, H. D. (1970). Role of ferredoxin and flavodoxin in the oxidative phosphorylation catalysed by cell-free preparations of the anaerobic *Desulfovibrio gigas*. *Bact. Proc.*, P75.

— (1971). Phosphorylation coupled to electron transfer between lactate and fumarate in cell-free extracts of the sulfate-reducing anaerobic *Desulfovibrio gigas*. *Bact. Proc.*, P155.

Bastin, E. S. (1926). A hypothesis of bacterial influence in the genesis of certain sulfide ores. *J. Geol.*, **34**, 773–92.

Basu, S. K. & Ghose, T. K. (1961). Bacterial sulphide production from sulphate-enriched spent distillery liquor II. *J. biochem. microbiol. Technol. Engng.*, **3**, 181–97.

Bauld, J., Chambers, L. A. & Skyring, G. W. (1979). Primary productivity, sulfate reduction and sulfur isotope fractionation in algal mats and sediments of Hamelin Pool, Shark Bay, W. A. *Aust. J. Mar. Freshwater Res.*, **30**, 753–64.

Baumann, A. & Denk, V. (1950). Zur Kenntnis der biologischen Sulfatreduktion. II. *Arch. Mikrobiol.*, **15**, 283–307.

Bavendamm, W. (1924). Die farblosen und rosen Schwefelbakterien. *Pflanzenforschung*, **2**, 1–156.

Beauchamp, R. S. A. (1953). Sulphates in African inland waters. *Nature, Lond.*, **171**, 769.

Beckwith, T. D. & Moser, J. R. (1932). The reduction of sulfur-containing compounds in wood pulp and paper manufacture. *J. Bact.*, **24**, 43–52.

Beerstecher, E. (1954). *Petroleum Microbiology*. Amsterdam: Elsevier.

Beijerinck, M. W. (1895). Über *Spirillum desulfuricans* als Ursache von Sulfatreduction. *Zentbl. Bakt. ParasitKde (abt. 2)*, **1**, 1–9, 49–59, 104–14.

Beijerinck, M. W. (1904). Phénomènes de réduction produits par les microbes. *Arch. néerl. Sci.*, **9**, 131–57.

Bell, G. R., Lee, J. P., Peck, H. D. & Le Gall, J. (1978). Reactivity of *Desulfovibrio gigas* hydrogenase towards artificial and natural electron donors. *Biochimie*, **60**, 315–20.

Bell, G. R., Le Gall, J. & Peck, H. D. (1974). Evidence for the periplasmic location of hydrogenase in *Desulfovibrio gigas*. *J. Bact.*, **120**, 994–7.

Bennett, E. D. & Bauerle, R. H. (1960). The sensitivities of mixed populations of bacteria to inhibitors. *Aust. J. biol. Sci.*, **13**, 142–9.

Bennett, E. O. (1956a). Control of bacterial spoilage of emulsion oils. *Soap & Chem. Specialities*, **32**, (10) 47 and (11) 46.

— (1956b). The role of sulfate-reducing bacteria in the deterioration of cutting emulsions. *Lubrication Engng.* **13**, 215–19.

Bernal, J. D. (1967). *The origin of life*. London: Weidenfeld & Nicholson.

Bianco, P., Fauque, G. & Haladjian, J. (1979). Electrode reaction of cytochrome c_3 from *Desulfovibrio vulgaris* Hildenborough and *Desulfovibrio desulfuricans* Norway. *Bioelectrochem. Bioenerget*, **6**, 385–91. *J. Electroanal. Chem.*, **104**, 385–91.

Biebl, F. & Pfennig, N. (1977). Growth of sulfate-reducing bacteria with sulfur as electron acceptor. *Arch. Microbiol.*, **112**, 115–17.

— (1978). Growth yields of green sulfur bacteria in mixed cultures with sulfur and sulfate-reducing bacteria. *Arch. Microbiol.*, **117**, 9–16.

Board, P. (1976). Anaerobic regulation of atmospheric oxygen. *Atmos. Environ.*, **10**, 339–42.

Boon, J. J., de Leeuw, J. W., van der Hoek, G. J. & Vosjan, J. H. (1977). Significance and taxonomic value of iso- and anteiso-monoenoic fatty acids and branched ß-hydroxy acids in *Desulfovibrio desulfuricans*. *J. Bact.*, **129**, 1183–91.

Boone, D. R. & Bryant, M. P. (1980). Propionate-degrading bacterium *Syntrophobacter wolinii* sp. nov., gen. nov. from methanogenic eco-systems. *Appl. env. Microbiol.*, **40**, 626–32.

Booth, G. H. (1964). Sulphur bacteria in relation to corrosion. *J. appl. Bact.*, **27**, 174–81.

Booth, G. H., Cooper, A. W. & Cooper, P. M. (1967). Criteria of soil aggressiveness towards buried metals. II. Assessment of various soils. *Br. J. Corrosion*, **2**, 109–15.

Booth, G. H., Cooper, A. W., Cooper, P. M. & Wakerley, D. S. (1967). Criteria of soil aggressiveness towards buried metals. I. Experimental methods. *Br. J. Corrosion* **2**, 104–8.

Booth, G. H., Cooper, A. W. & Tiller, A. K. (1967). Criteria of soil aggressiveness towards buried metals. III. Verification of predicted behaviour of selected soils. *Br. J. Corrosion*, **2**, 116–18.

Booth, G. H., Elford, L. & Wakerley, D. S. (1968). Corrosion of mild steel by sulphate-reducing bacteria: an alternative mechanism. *Br. J. Corrosion*, **3**, 242–5.

Booth, G. H. & Mercer, S. J. (1963). Resistance to copper of some oxidizing and reducing bacteria. *Nature, Lond.*, **199**, 622.

Booth, G. H., Shinn, P. M. & Wakerley, D. S. (1965). The influence of various strains of actively growing sulphate-reducing bacteria on the anaerobic corrosion of mild steel. *C.R. Congr. Intern. Corrosion Marine et Salissures*, pp. 363–71.

Brandis, A. & Thauer, R. K. (1981). Growth of *Desulfovibrio* species of hydrogen and sulphate as sole energy source. *J. gen. Microbiol.*, **126**, 249–52.

Brisou, J. & De Rautlin de la Roy, Y. (1965). Le role des bactéries aerobies sulfhydrogènes dans la corrosion des métaux. *Congr. Intern. Corrosion Marine 1964, Cannes*, pp. 373–5.

Broda, E. (1975). *The evolution of bioenergetic processes.* Oxford: Pergamon Press.

Bruschi, M. (1979). Amino acid sequence of *Desulfovibrio gigas* ferredoxin: revisions. *Biochem. Biophys. Res. Commun.*, **91**, 623–8.

— (1981). The primary structure of the tetrahaem cytochrome c_3 from *Desulfovibrio desulfuricans* (strain Norway 4). Description of a new class of low-potential cytochrome *c*. *Biochim. biophys. Acta*, **671**, 219–26.

Bruschi, M. & Hatchikian, E. C. (1982). Non-heme iron proteins of *Desulfovibrio*: the primary structure of ferredoxin I from *Desulfovibrio africanus*. *Biochimie*, **64**, 503–7.

Bruschi, M., Hatchikian, E. C., Bonicel, J., Bovier-Lapierre, G. & Couchoud, P. (1977). The *N*-terminal sequence of superoxide dismutase from the strict anaerobe *Desulfovibrio desulfuricans*. *FEBS Lett.*, **76**, 121–4.

— Bruschi, M., Hatchikian, E. C., Le Gall, J., Moura, J. J. G. & Xavier, A. V. (1976). Purification, characterisation and biological activity of three forms of ferredoxin from the sulfate-reducing bacterium *Desulfovibrio gigas*. *Biochim. biophys. Acta*, **449**, 275–84.

Bruschi, M., Le Gall, J. & Dus, K. (1970). C-type cytochromes of *Desulfovibrio vulgaris* amino acid composition and end groups of cytochrome C_{553}. *Biochem. biophys. Res. Commun.*, **38**, 607–16.

Bruschi, M., Le Gall, J., Hatchikian, E. C. & Dubordieu, M. (1969). Cristallisation et proprietés d'un cytochrome intervenant dans la réduction du thiosulfate par *Desulfovibrio gigas*. *Bull. Soc. Fr. Physiol. vég.*, **15**, 381–90.

Bruschi, M., Moura, I., Le Gall, J., Xavier, A. V. & Sieker, L. C. (1979). The amino-acid sequence of desulforedoxin, a new type of non-heme iron protein from *Desulfovibrio gigas*. *Biochem. biophys. Res. Commun.*, **90**, 596–605.

Bryant, M. P., Campbell, L. L., Reddy, C. A. & Crabill, M. R. (1977). Growth of Desulfovibrio in lactate or ethanol media low in sulfate in association with H_2-utilizing methanogenic bacteria. *Appl. environ. Microbiol*, **33**, 1162–9.

Buller, C. S. & Akagi, J. M. (1964). Hydrogenase of Coleman's sulfate-reducing bacterium. *J. Bact.*, **88**, 440–3.

Bunker, H. J. (1936). *A review of the physiology and biochemistry of the sulphur bacteria*. London: HM Stationery Office.

Bunker, H. J. (1942). The control of bacterial sulphate reduction by regulation of hydrogen ion concentration. *Proc. Soc. agric. Bact.*, 8–10. (*Abs*).

Burgess, S. G. & Wood, L. B. (1961). Pilot plant studies in production of sulphur from sulphate-enriched sewage sludge. *J. Sci. Fd Agric.*, **4**, 326–35.

Burton, K. (1955). The free energy change associated with the hydrolysis of the thiol ester bond of acetyl co-enzyme A. *Biochem. J.*, **59**, 44–6.

Burton, K. & Krebs, H. A. (1953). The free energy changes associated with the individual steps of the tricarboxylic acid cycle, glycolysis and alcoholic fermentation and with the hydrolysis of the pyrophosphate groups of adenosine triphosphate. *Biochem. J.*, **54**, 94–107.

Butlin, K. R. (1949). Some malodorous activities of sulphate-reducing bacteria. *Proc. Soc. appl. Bact.*, **2**, 39–42.

Butlin, K. R. & Adams, M. E. (1947). Autotrophic growth of sulphate-reducing bacteria. *Nature, Lond.*, **160**, 154.

Butlin, K. R., Adams, M. E. & Thomas, M. (1949). The isolation and cultivation of sulphate-reducing bacteria. *J. gen. Microbiol.*, *3*, 46–59.

Butlin, K. R. & Postgate, J. R. (1954*a*). The microbiological formation of sulphur in Cyrenaican lakes. In *Biology of deserts*, ed Cloudsley-Thompson, J., pp. 112–22. London: Inst. Biol.

— (1954*b*). The economic importance of autotrophic micro-organisms. In *Autotrophic micro-organisms*. ed. Fry, B. A. & Peel, J. L. *Symp. Soc. gen. Microbiol*, **4**, 271–305.

Butlin, K. R., Selwyn, S. C. & Wakerley, D. S. (1956). Sulphide production from sulphate-enriched sewage sludges. *J. appl. Bact.*, **19**, 3–15.

— (1960). Microbial sulphide production from sulphate-enriched sewage sludge. *J. appl. Bact.*, **23**, 158–68.

Cabasso, V. & Roussel, H. (1942). Essai d'explication du phénomène dit 'des eaux rouges' du Lac de Tunis. *Arch. Inst. Pasteur Tunis*, **31**, 203–11.

Callender, S. E. & Roberts, E. R. (1961). Studies in the biological fixation of nitrogen. XI. Reduction of azo and other unsaturated linkages by hydrogenase. *Biochim. biophys. Acta*, **54**, 92–101.

Cammack, R., Patil, D., Aguirre, R. & Hatchikian, E. C. (1982). Redox properties of the ESR-detectable nickel in hydrogenase from *Desulfovibrio gigas*. *FEBS Lett.*, **142**, 289–92.

Cammack, R., Rao, K. K., Hall, D. O., Moura, J. J. G., Xavier, A. V., Bruschi, M., Le Gall, J., Deville, A. & Goyda, J. P. (1977). Spectroscopic studies of the oxidation–reduction properties of three forms of ferredoxin from *Desulphovibrio gigas. Biochim. biophys. Acta*, **490**, 311–21.

Campbell, L. L., Frank, H. A. & Hall, E. R. (1956). Studies on thermophilic sulphate reducing bacteria. I. Identification of *Sporovibrio desulfuricans* as *Clostridium nigrificans*. *J. Bact.*, **73**, 516–21.

Campbell, L. L. & Postgate, J. R. (1965). Classification of the spore-forming sulfate-reducing bacteria. *Bact. Rev.*, **29**, 359–63.

— (1969). Revision of the holotype strain of *Desulfotomaculum ruminis* (Coleman) Campbell & Postgate. *Int. J. syst. Bact.*, **19**, 139–40.

Cappenberg, T. E. (1974*a*). Ecological observations on heterotrophic, methane-oxidizing and sulfate-reducing bacteria in a pond. *Hydrobiologia*, **40**, 471–85.

— (1974*b*). Interrelations between sulfate-reducing and methane-producing bacteria in bottom deposits of a fresh water lake. I. Field observations. *Antonie van Leeuwenhoek*, **40**, 285–95.

— (1974*c*). Interrelations between sulfate-reducing and methane-producing bacteria in bottom deposits of a fresh water lake. II. Inhibition experiments. *Antonie van Leeuwenhoek*, **40**, 297–306.

— (1975*a*). Relationships between sulfate-reducing and methane-producing bacteria. *Pl. Soil*, **43**, 125–39.

— (1975*b*). A study of mixed continuous cultures of sulfate-reducing and methane-producing bacteria. *Microb. Ecol*, **2**, 60–72.

Cavanaugh, C. M., Gardiner, S. L., Jones, M. L., Jannasch, H. W. & Waterbury, J. B. (1981). Prokaryotic cells in the hydrothermal vent tube worm *Riftia pachyptila* Jones: possible chemoautotrophic symbionts. *Science*, **213**, 340–2.

Chambers, L. A. & Trudinger, P. A. (1975). Are thiosulfate and trithionate intermediates in dissimilatory sulfate reduction? *J. Bact*, **123**, 36–40; *erratum* (1976), J. Bact., **125**, 387.

Chemistry Research 1952, London: HM Stationery Office.

— 1953, London: HM Stationery Office.

— 1956, London: HM Stationery Office.

— 1957, London: HM Stationery Office.

Cinquina, C. L. (1968). Isolation of tocopherol from *Desulfotomaculum nigrificans*. *J. Bact.*, **95**, 2436–8.

Clark, B. C. (1979). Solar driven chemical energy source for a Martian biota. *Origins of Life*, **9**, 241–50.

— (1981). Sulfur: Fountainhead of life in the Universe? In *Life in the Universe*, ed Billingham, J., pp. 47–60. Cambridge, Mass.: MIT Press.

Clark, B. C. & Baird, A. K. (1979). Is the Martian lithosphere sulfur rich? *J. Geophys. Res.*, **84**, 8395–403.

Clark, B. C. & van Hart, D. C. (1981). The salts of Mars. *Icarus*, **45**, 370–8.

Coleman, G. S. (1960). A sulphate-reducing bacterium from sheep rumen. *J. gen. Microbiol*, **22**, 423–36.

Copenhagen, W. J. (1934). Occurrence of sulphides in certain areas of the sea bottom on the South African coast. *Un. S. Afr. Comm. Ind., Fish Mar. Biol. Sur. Div.*, Investigational Report No. 3

— (1953). The periodic mortality of fish in the Walvis region. *Un. S. Afr. Dept. Comm. Ind., Fish Div., Investigational Report No. 14*.

References

Costello, D., King, R. A. & Miller, J. D. A. (1970). Influence of sodium chloride on inhibition of *Desulfovibrio* by a surfactant. *Arch. Mikrobiol*, **71**, 196–8.

Crombie, D. J., Moodie, G. J. & Thomas, J. D. R. (1980). Corrosion of iron by sulphate-reducing bacteria. *Chem. Ind.*, 500–04.

Curtis, W. & Ordal, E. J. (1954). Hydrogenase and hydrogen metabolism in *Micrococcus aerogenes*. *J. Bact.*, **68**, 351–61.

Daly, D. J. & Anderson, K. E. (1966). Threonine metabolism of *Desulfovibrio desulfuricans*. *Bact. Proc.*, P201.

Davis, J. B. (1967). *Petroleum Microbiology*. Amsterdam: Elsevier.

Davis, J. B. & Yarbrough, J. F. (1966). Anaerobic oxidation of hydrocarbons by *Desulfovibrio desulfuricans*. *Chem. Geol.*, **1**, 137–44.

Delden, A. van (1903). Beitrag zur Kenntris der Sulfatreduktion durch Bakterien. *Zentbl. Bakt. ParasitKde (abt. 2)*, **11**, 81–94, 113–19.

DerVartanian, D. V. & Le Gall, J. (1971). Electron paramagnetic resonance studies on the reaction of exogenous ligands with cytochrome c_3 from *Desulfovibrio vulgaris*. *Biochim. biophys. Acta*, **243**, 53–65.

DerVartanian, D. V. & Le Gall, J. (1974). A mononuclear electron transfer chain: structure and function of cytochrome c_3. *Biochim. biophys. Acta*, **346**, 79–99.

DerVartanian, D. V., Xavier, A. V. & Le Gall, J. (1978). EPR determination of the oxidation reduction potentials of the hemes in cytochrome c_3 from *Desulfovibrio vulgaris*. *Biochimie*, **60**, 321–5.

Desrochers, R. & Fredette, V. (1960). Etude d'une population de bactéries réductrices du soufre. *Can. J. Microbiol.*, **6**, 349–54.

Dicker, H. J. & Smith, D. W. (1980). Enumeration and relative importance of acetylene-reducing (nitrogen-fixing) bacteria in a Delaware salt marsh. *Appl. env. Microbiol*, **39**, 1019–25.

Dickerson, R. E., Timkovich, R. & Almassy, R. J. (1976). The cytochrome fold and the evolution of bacterial energy metabolism. *J. mol. biol.*, **100**, 473–91.

Dittbrenner, S., Chowdury, A. A. & Gottschalk, G. (1969). The stereospecificity of the (R)-citrate synthase in the presence of *p*-chloromercuribenzoate. *Biochem. biophys. Res. Commun.*, **36**, 802–8.

Dobson, C. M., Hoyle, N. J., Geraldes, C. F., Wright, P. E., Williams, R. J. P., Bruschi, M. & Le Gall, J. (1974). Outline structure of cytochrome c_3 and consideration of its properties. *Nature, Lond.*, **249**, 425–9.

Domka, F. & Gasiowek, J. (1975). Role of micro-organisms in the reduction of sulfates to form sulfur deposits in the Carpathian region. *Przegl. Geol.*, **23**, 61–5 (in Polish).

Domka, F., Gasiowek, J. & Klemm, A. (1977). Processing of sulfate wastes using municipal sewage. *Gaz. Wodka. Tech. Sanit*, **51**, 179–80 (in Polish).

Domka, F. & Szulczynski, M. (1981). Utilisation de certains analogues du glycerophosphate par une bactérie sulfato-réductrice *Desulfovibrio vulgaris*. *Ann. Microbiol. (Inst. Pasteur)*, **132**, 107–11.

Dommergues, Y., Combremont, R., Beck, G. & Ollat, C. (1969). Note préliminaire concernant la sulfato-réduction rhizospherique dans un sol salin Tunisien. *Rev. Ecol. Biol. Sol*, T. VI, **2**, 115–29.

Donelly, L. S. & Busta, F. F. (1980). Heat resistance of *Desulfotomaculum nigrificans* spores in soyprotein infant formula preparations. *Appl. env. Microbiol.*, **40**, 721–5.

Dostalek, M. (1961). Bacterial release of oil. III. *Folia Microbiol., Praha*, **6**, 10–17.

Dostalek, M. & Kvet, R. (1964). Utilization of the osmo-tolerance of sulphate reducing bacteria in study of the genesis of subterranean waters. *Folia Microbiol., Praha*, **9**, 103–14.

Dostalek, M. & Spurny, M. (1958). Bacterial release of oil. I. *Folia Microbiol., Praha*, **4**, 166–72.

Drake, H. L. & Akagi, J. M. (1976). Purification of a unique bisulfite-reducing enzyme from *Desulfovibrio vulgaris*. *Biochem. biophys. Res. Commun.*, **71**, 1214–19.

— (1977). Bisulfite reductase of *Desulfovibrio vulgaris*: explanation for product formation. *J. Bact.*, **132**, 139–43.

— (1978). The dissimilatory reduction of bisulfite by *Desulfovibrio vulgaris*. *J. Bact.*, **136**, 916–23.

Drucker, H. & Campbell, L. L. (1969). Electrophoretic and immunological difference between the cytochrome c_3 of *Desulfovibrio desulfuricans* and that of *Desulfovibrio vulgaris*. *J. Bact.*, **100**, 358–64.

Drucker, H., Trousil, E. B. & Campbell, L. L. (1970). Purification and properties of cytochrome c_3 of *Desulfovibrio salexigens*. *Biochemistry*, **9**, 3395–400.

Drummond, J. P. M. & Postgate, J. R. (1955). A note on the enumeration of sulphate-reducing bacteria in polluted water and on their inhibition by chromate. *J. appl. Bact.*, **18**, 307–11.

Dubourdieu, M. & Fox, J. L. (1977). Amino acid sequence of *Desulfovibrio vulgaris* flavodoxin. *J. biol. Chem.*, **252**, 1453–63.

Elion, L. (1925). A thermophilic sulphate-reducing bacterium. *Zentbl. Bakt. ParasitKde, (abt. 2)*, **63**, 58–67.

Ellis, D. (1932). Dilution of sewage in the sea. *J. Roy. Tech. Coll. (Glasgow)*, **2**, 698–707.

Fauque, G. D., Barton, L. L. & Le Gall, J. (1980). Oxidative phosphorylation linked to dissimilatory reduction of elemental sulphur by *Desulfovibrio*. In *Sulphur in biology. Ciba Foundation Symp. 72*, pp. 71–86.

Fee, J. A. (1982). Is superoxide important in oxygen poisoning? *Trends in Biochem. Sci.*, **7**, 84–6.

Fenchel, T. M. & Jørgensen, B. B. (1977). Detritus food chains of aquatic ecosystems: the role of bacteria. *Adv. microbial Ecol.*, **1**, 1–58; see especially p. 29.

Fenchel, T. M. & Riedl, R. J. (1970). The sulfide system: a new biotic community underneath the oxidized layer of marine sand bottoms. *Mar. Biol.*, **7** 255–68.

Findley, J. E. & Akagi, J. M. (1968). Lysis of *Desulfovibrio vulgaris* by ethylenediaminetetraacetic acid and lysozyme. *J. Bact.*, **96**, 1427–8.

— (1969). Evidence for thiosulfate formation during sulfite reduction by *Desulfovibrio vulgaris*. *Biochem. biophys. Res. Commun.*, **36**, 266–71.

— (1970). Role of thiosulfate in bisulfite reduction as catalysed by *Desulfovibrio vulgaris*. *J. Bact.*, **103**, 741–4.

Ford, H. W. (1965), Bacterial metabolites that affect citrus root survival in soils subject to flooding. *Amer. Soc. hort. Sci.*, **86**, 205–11.

Foster, J. W. (1962). Hydrocarbons as substrates for micro-organisms. *Antonie van Leeuwenhoek*, **28**, 241–74.

Fox, G. E., Stackebrandt, E., Hespell, R. B., Gibson, J., Maniloff, J., Dyer, T. A., Wolfe, R. S., Balch, W. E., Tanner, R. S., Magrum, L. J., Zablen, L. B., Blakemore, R., Gupta, R., Bonen, L., Lewis, B. J., Stahl, D. A., Luehrsen, K. R., Chen, K. N. & Woese, C. R. (1980). The phylogeny of prokaryotes. *Science*, **209**, 457–63.

References

Freke, A. M. & Tate, D. (1961). The formation of magnetic iron sulphide by bacterial reduction of iron solutions. *J. Biochem. Microbiol. Tech. Eng.*, **3**, 29–39.

Furusaka, C. (1961). Sulphate transport and metabolism by *Desulphovibrio desulphuricans*. *Nature, Lond.*, **192**, 427–9.

Garland, P. B. (1977). Energy transduction and transmission in microbial systems. In *Microbial energetics*, ed. Haddock, B. A. & Hamilton, W. A. *Symp. Soc. gen. Microbiol.*, **27**, 1–22.

Genovese, S. (1963). The distribution of the H_2S in the Lake of Faro (Messina) with particular regard to the presence of 'red water'. In *Marine Microbiology* ed. Oppenheimer, C. H., pp. 194–204. Springfield, Ill., USA: C. C. Thomas.

Germano, G. J. & Anderson, K. E. (1967). Biosynthesis of alanine by *Desulfovibrio desulfuricans*. *Bact. Proc.*, P131.

— (1968). Purification and properties of L-alanine dehydrogenase from *Desulfovibrio desulfuricans*. *J. Bact.*, **96**, 55–60.

— (1969). Serine biosynthesis in *Desulfovibrio desulfuricans*. *J. Bact.*, **99**, 893–4.

Gevertz, D., Amelunxen, R. & Akagi, J. M. (1980). Cysteine synthesis by *Desulfovibrio vulgaris* extracts. *J. Bact.*, **141**, 1460–2.

Ghose, T. K. & Basu, S. K. (1961). Bacterial sulphide production from sulphate enriched spent distillery liquor I. *Folia Microbiol., Praha*, **6**, 335–41.

Ghose, T. K., Mukkerjee, S. K. & Basu, S. K. (1964). Bacterial sulphide production from sulphate-enriched spent distillery slop. III. *Biotechnol. Bioengng.*, **6**, 285–97.

Glick, B. R., Martin, W. G. & Martin, S. M. (1980). Purification and properties of the periplasmic hydrogenase from *Desulfovibrio desulfuricans*. *Can. J. Microbiol.*, **26**, 1214–23.

Gottschalk, G. & Andreesen, J. R. (1979). Energy metabolism in anaerobes. *Int. Rev. Biochem.*, **21**, 85–115.

Gottschalk, G. & Barker, H. A. (1967). Presence and stereospecificity of citrate synthase in anaerobic bacteria. *Biochemistry*, **6**, 1027–34.

Granat, L., Hallberg, R. O. & Rodhe, H. (1976). The global sulphur cycle. In *Nitrogen, phosphorus and sulphur-global cycles*, ed. Svensson, B. H. & Soderlund, R. *Ecol. Bull (Stockholm)*, **20**, 89–134.

Gray, B. H. (1977). Rejection of *Chloropseudomonas ethylica* as a *nomina rejicienda*: request for an opinion. *Int. J. syst. Bact.*, **27**, 168.

Gray, B. H., Fowler, C. F., Nugent, N. A., Rigopoulos, N. & Fuller, R. C. (1973). Reevaluation of *Chloropseudomonas ethylica* strain 2-K. *Int. J. Syst. Bact*, **23**, 256–64.

Greathouse, G. A. & Wessel, C. J. (1954). *Deterioration of materials*. New York: Reinhold.

Grossman, J. P. & Postgate, J. R. (1953). Cultivation of sulphate-reducing bacteria. *Nature, Lond.*, **171**, 600–2.

— (1955). The metabolism of malate and certain other compounds by *Desulphovibrio desulphuricans*. *J. gen. Microbiol.*, **12**, 429–45.

Guerlesquin, F., Bovier-Lapierre, G. & Bruschi, M. (1982). Purification and characterization of cytochrome c_3 (Mr 26,000) isolated from *Desulfovibrio desulfuricans* Norway strain. *Biochem. biophys. Res. Commun.*, **105**, 530–8.

Guerlesquin, F., Moura, J. J. G. & Cammack, R. (1982). Iron–sulphur cluster composition and redox properties of two ferredoxins from *Desulfovibrio desulfuricans* Norway strain. *Biochim. biophys. Acta*, **679**, 422–7.

Gunkel, W. & Oppenheimer, C. H. (1963). Experiments regarding the sulfide formation in sediments of the Texas Gulf coast. In *Marine Microbiology*, ed. Oppenheimer, C. H. pp. 674–84. Springfield, Ill., USA: Thomas.

Guttierez, J. (1953). Numbers and characteristics of lactate-utilizing organisms in the rumen of cattle. *J. Bact.*, **66**, 123–8.

Guynes, G. J. & Bennett, E. O. (1959). Bacterial deterioration of emulsion oils. I. Relationship between aerobes and sulfate-reducing bacteria in deterioration. *Appl. Microbiol.*, **7**, 117–21.

Haines, T. H., Henry, J. M. & Block, R. J. (1960). The sulfur metabolism of insects. V. The ability of insects to use sulfate in the synthesis of methionine. *Contrib. Boyce Thompson Inst.*, **20**, 363–4.

Hall, H. M., Prince, R. H. & Cammack, R. (1979). EPR spectroscopy of the iron-sulphur cluster and sirohaem in the dissimilatory sulphite reductase (desulphoviridin) from *Desulphovibrio gigas*. *Biochim. biophys. Acta*, **581**, 27–33.

Han, J. & Calvin, M. (1969). Hydrocarbon distribution of algae and bacteria, and microbiological activity in sediments. *Proc. natn. Acad. Sci. USA*, **64**, 436–43.

Handley, J., Adams, V. & Akagi, J. M. (1973). Morphology of bacteriophage-like particles from *Desulfovibrio vulgaris*. *J. Bact.*, **115**, 1205–7.

Hardy, J. A. & Hamilton, W. A. (1981). The oxygen tolerance of sulfate-reducing bacteria isolated from North Sea waters. *Current Microbiology*, **6**, 259–62.

Harrison, A. G. & Thode, H. G. (1958). Mechanism of the bacterial reduction of sulphate from isotope fractionation studies. *Trans. Faraday Soc.*, **54**, 84–92.

Haschke, R. H. & Campbell, L. L. (1971). Purification and properties of a hydrogenase from *Desulfovibrio vulgaris*. *J. Bact.*, **105**, 249–58.

Haser, R., Pierrot, M., Frey, M., Payan, F., Astier, J.P., Bruschi, M. & Le Gall, J. (1979). Structure and sequence of the multihaem cytochrome c_3. *Nature, Lond.*, **282**, 806–10.

Hatchikian, E. C. (1981). A cobalt porphyrin containing protein reducible by hydrogenase isolated from *Desulfovibrio desulfuricans* (Norway). *Biochem. biophys. Res. Commun.*, **103**, 521–30.

Hatchikian, E. C. & Bruschi, M. (1979). Isolation and characterization of a molybdenum iron-sulfur protein from *Desulfovibrio africanus*. *Biochem. biophys. Res. Commun.*, **86**, 725–34.

Hatchikian, E. C., Bruschi, M. & Le Gall, J. (1978). Characterization of the periplasmic hydrogenase from *Desulfovibrio gigas*. In *Hydrogenases: their catalytic activity, structure and function*, ed. Schlegel, H. G. & Schneider, K., pp. 93–106. Göttingen: Golze.

Hatchikian, E. C., Le Gall, J., Bruschi, M. & Dubourdieu, M. (1972). Regulation of the reduction of sulfite and thiosulfate by ferredoxin, flavodoxin and cytochrome cc'_3 in extracts of the sulfate reducer *Desulfovibrio gigas*. *Biochim. biophys. Acta*, **258**, 701–8.

Hatchikian, E. C., Jones, H. E. & Bruschi, M. (1979). Isolation and characterization of a rubredoxin and two ferredoxins from *Desulfovibrio africanus*. *Biochim. biophys. Acta*, **548**, 471–83.

Hatchikian, E. C. & Le Gall, J. (1970*a*). Etude du métabolisme des acides dicarboxyliques et du pyruvate chez les bactéries sulfato-réductrices: I. Etude de l'oxydation enzymatique de fumarate en acetate. *Annls. Inst. Pasteur, Paris*, **118**, 125–42.

— (1970*b*). Etude du métabolisme des acides dicarboxyliques et du pyruvate chez les bactéries sulfato-réductrices: II. Transport des electrons: accepteurs finaux. *Annls. Inst. Pasteur, Paris*, **118**, 288–301.

— (1972). Evidence for the presence of a b-type cytochrome in the sulfate-reducing bacterium *Desulfovibrio gigas* and its role in the reduction of fumarate by molecular hydrogen. *Biochim. biophys. Acta*, **267**, 479–84.

Hatchikian, E. C., Le Gall, J. & Bell, G. R. (1977). Significance of superoxide dismutase and catalase activities in the strict anaerobes, sulfate-reducing bacteria. In *Superoxide and superoxide dismutases*, ed. Michelson, A. M., McCord, J. M. & Fridovich, I., pp. 159–72. London: Academic Press.

Hatchikian, E. C. & Monson, P. (1980). Highly active immobilized hydrogenase from *Desulfovibrio gigas*. *Biochem. biophys. Res. Commun.*, **92**, 1091–6.

Hayes, F. R. & Coffin, C. C. (1951). Radioactive phosphorus and exchange of lake nutrients. *Endeavour*, **10**, 78–81.

Hayward, H. R. (1960). Anaerobic degradation of choline, III. *J. biol. Chem.*, **235**, 3592–6.

Hayward, H. R. & Stadtman, T. C. (1959). Anaerobic degradation of choline, I. *J. Bact*, **78**, 557–61.

— (1960). Anaerobic degradation of choline, II. *J. biol. Chem.*, **235**, 538–43.

Herbert, B. N. (1976). The effect of hydrostatic pressure on bacteria in seawater intended for injection into oil formations. *J. appl. Bact.*, **41**, xii.

Herbert, R. A. (1975). Heterotrophic nitrogen fixation in shallow estuarine environments. *J. exp. mar. Biol. Ecol.*, **18**, 215–25.

Hesse, P. R. (1956). Sulphate deficiency in Lake Victoria. *Nature, Lond.*, **177**, 389–90.

Hewitt, J. & Morris, J. G. (1975). Superoxide dismutase in some obligately anaerobic bacteria. *FEBS Lett*, **50**, 315–18.

Higuchi, Y., Bando, S., Kusunoki, M., Matsuura, Y., Yasuoka, N., Kakudo, M., Yamanaka, T., Yagi, T. & Inokuchi, H. (1981). The structure of cytochrome c_3 from *Desulfovibrio vulgaris* Miyazaki at 2.5 Å resolution. *J. Biochem*, **89**, 1659–62.

Hill, E. C. (1975). Biodeterioration of petroleum products. In *Microbial aspects of the deterioration of materials* SAB Tech. Ser. 9, ed Lovelock, D. W. & Gilbert, R. J., pp. 127–36. London: Academic Press.

Hill, S., Drozd, J. W. & Postgate, J. R. (1972). Environmental effects on the growth of nitrogen-fixing bacteria. *J. appl. Chem. Biotechnol.*, **22**, 541–58.

Hines, M. E. & Buck, J. D. (1982). Distribution of methanogenic and sulfate-reducing bacteria in near-shore marine sediments. *Appl. env. Microbiol.*, **43**, 447–53.

Hobson, P. N., Summers, R., Postgate, J. R. & Ware, D. (1973). Nitrogen fixation in the rumen of the living sheep. *J. gen. Microbiol*, **77**, 225–6.

Hodgman, C. D. (1949). *Handbook of chemistry and physics*. Cleveland: Chemical Rubber Publn. Co.

Howard, B. H. & Hungate, R. E. (1976). *Desulfovibrio* of the sheep rumen. *Appl. env. Microbiol.*, **32**, 598–602.

Howarth, R. W. & Teal, J. M. (1979). Sulfate reduction in a New England salt marsh. *Limnol. Oceanogr.*, **24**, 999–1013.

Huisingh, J., McNiell, J. J. & Matrone, G. (1974). Sulfate reduction by a *Desulfovibrio* species isolated from sheep rumen. *Appl. Microbiol.*, **28**, 489–97.

Huynh, B. H., Moura, J. J. G., Moura, I., Kent, T. A., Le Gall, J., Xavier, A. V., & Munck, E. (1980). Evidence for a three-iron center in a ferredoxin from *Desulfovibrio gigas*. Mossbauer and EPR studies. *J. biol. Chem.*, **255**, 3242–44.

Hvid-Hansen, N. (1951). Sulphate-reducing and hydrocarbon-producing bacteria in ground water. *Acta path. microbiol. scand.*, **29**, 315–35.

Iizuka, H., Okahazi, H. & Seto, N. (1969). A new sulfate-reducing bacterium isolated from Antarctica. *J. gen. appl. Microbiol., Tokyo*, **15**, 11–18.

Isenberg, D. L. & Bennett, E. O. (1959). Bacterial deterioration of emulsion oils. II. Nature of the relationship between aerobes and sulfate-reducing bacteria. *Appl. Microbiol.*, **7**, 121–5.

Ishimoto, M. (1959). Sulfate reduction in cell-free extracts of Desulfovibrio. *J. Biochem., Tokyo*, **46**, 105–6.

Ishimoto, M. & Koyama, J. (1957). Biochemical studies on sulfate-reducing bacteria. VI. Separation of hydrogenase and thiosulfate reductase and partial purification of cytochrome and green pigment. *J. Biochem., Tokyo*, **44**, 233–42.

Ishimoto, M., Koyama, J. & Nagai, Y. (1954a). Role of a cytochrome in thiosulfate reduction by a sulfate reducing bacterium. *Seikagaku zasshi*, **26**, 303.

— (1954b). Biochemical studies on sulfate-reducing bacteria. IV. The cytochrome system of sulfate-reducing bacteria. *J. Biochem., Tokyo*, **41**, 763–70.

— (1955). Biochemical studies on sulfate-reducing bacteria. V. Reduction of thiosulfate by cell-free extract. *J. Biochem., Tokyo*, **42**, 41–53.

Ivanov, M. V. (1964). Microbiological processes in the formation of sulfur deposits. *US Dept. Agric. and National Sci. Foundation*, Jerusalem: Israel Program for Scientific Translations (1968).

Iverson, W. P. (1967). Disulfur monoxide; production by Desulfovibrio. *Science*, **156**, 1112–14.

— (1972). Biological corrosion. *Adv. Corrosion Sci. Technol.*, **2** 1–42.

— (1974). Microbial corrosion of iron. In *Microbial iron metabolism*, ed. Nielands, J. B. pp. 475–513. Academic Press: New York, London, San Francisco.

Iya, K. K. & Sreenivasaya, M. (1945). Studies in the sulphur formation at Kona, Masulipatam – part II. *Curr. Sci.*, **14**, 267–9.

Jacq, V. & Dommergues, Y. (1970). Sulfato-réduction rhizosphérique et spermosphérique: influence de la densité apparente du sol. *Seances Acad. Agric. France*, pp. 511–19.

— (1971). Sulfato-réduction spermosphérique. *Annls. Inst. Pasteur, Paris* **121**, 199–206.

Jacq, V. A. & Fortuner, R. (1980). Biological control of rice nematodes using sulphate-reducing bacteria: laboratory and field trials. *Abs. 2nd Intern. Symp. Microbial Ecol.* (Univ. of Warwick, UK), p. 203.

Jankowski, G. J. & ZoBell, C. E. (1944). Hydrocarbon production by sulfate-reducing bacteria. *J. Bact.*, **47**, 447.

Jannasch, H. W. (1957). Die bakterielle Rotfärbung der Salzseen des Wadi Natrun (Ägypten). *Arch. Hydrobiol.*, **53**, 425–33.

Jannasch, H. J., Wirsen, C. O. & Taylor, C. D. (1976). Undecompressed microbial populations from the deep sea. *Appl. environ. Microbiol.*, **32**, 360–7.

Jensen, M. L. (ed.) (1962). *Biogeochemistry of sulphur isotopes*. Newhaven: Yale Univ. Press.

Johnson, M. K., Hare, J. W., Spiro, T. G., Moura, J. J. G., Xavier, A. V. & Le Gall, J. (1981). Resonance Raman spectra of three-iron centers in ferredoxins from Desulfovibrio gigas. *J. biol. Chem.*, **256**, 9806–8.

Jones, C. W. (1981). Unity and diversity in bacterial energy conservation. In *Contemporary Microbial Ecology*, ed. Ellwood, D. C., Hedger, J. N., Latham, M. J., Lynch, J. M. & Slater, J. H., pp. 193–213. Academic Press: London.

Jones, G. E. & Starkey, R. L. (1957). Fractionation of stable isotopes of sulfur by microorganisms and their role in deposition of native sulfur. *Appl. Microbiol.*, **5**, 111–18.

Jones, H. E. (1971a). A re-examination of *Desulfovibrio africanus*. *Arch. Mikrobiol.*, **80**, 78–86.
— (1971b). Sulfate-reducing bacterium with unusual morphology and pigment content. *J. Bact.*, **106**, 339–46.
— (1972). Cytochromes and other pigments of dissimilatory sulphate-reducing bacteria. *Arch. Mikrobiol*, **84**, 207–24.
Jones, H. E. & Skyring, G. W. (1974). Reduction of sulphite to sulphide catalysed by desulfoviridin from *Desulfovibrio gigas*. *Aust. J. biol. Sci.*, **27**, 7–14.
— (1975). Effect of enzymic assay conditions on sulfite reduction catalysed by desulfoviridin from *Desulfovibrio gigas*. *Biochim. biophys. Acta*, **77**, 52–60.
Jørgensen, B. B. (1977). The sulfur cycle of a coastal marine sediment (Limfjorden, Denmark). *Limnol. Oceanogr.*, **22**, 814–32.
— (1978a). A comparison of methods for the quantification of bacterial sulfate reduction in coastal marine sediments. I. Measurement with radiotracer techniques. *Geomicrobiol. J.*, **1** 11–28.
— (1978b). A comparison of methods for the quantification of bacterial sulfate-reduction in coastal marine sediments. II. Calculation from mathematical models. *Geomicrobiol. J.*, **1**, 29–48.
— (1978c). A comparison of methods for the quantification of bacterial sulfate reduction in coastal marine sediments. III. Estimation from chemical and bacteriological field data. *Geomicrobiol. J.*, **1**, 49–64.
— (1982a). Ecology of the bacteria of the sulphur cycle with special reference of anoxic–oxic interface environments. *Phil. Trans. R. Soc. Lond. B.*., **298**, 543–61.
— (1982b). Mineralization of organic matter in the sea bed – the role of sulphate reduction. *Nature, Lond.*, **296**, 643–5.
— (1984). The microbial sulfur cycle. In *Microbial geochemistry* (ed. Krumbein, W. E.), pp. 00–00. Oxford: Blackwell.
Jørgensen, B. B. & Cohen, Y. (1977). Solar Lake (Sinai). 5. The sulfur cycle of the benthic cyanobacterial mats. *Limnol. Oceanogr.*, **22**, 657–66.
Kadota, H. & Miyoshi, H. (1964). The role of organic matter in the production of sulfides by sulfate-reducing bacteria in marine and estuarine sediments. *Bull. Res. Inst. Fd Sci. Kyoto Univ.*, No **27**, 9–29.
Kamen, M. D. & Horio, T. (1970). Bacterial cytochromes. I. Structural aspects. *A. Rev. Biochem.*, **39**, 673–700.
Kaplan, I. R. (1956). Evidence of microbiological activity in some of the geothermal regions of New Zealand. *NZ J. Sci. Technol.*, **37**, 639–62.
Kaplan, I. R. & Rittenberg, S. C. (1962). The microbiological fractionation of sulfur isotopes. In *Biogeochemistry of sulfur isotopes*, ed Jensen, M. L., pp. 80–93. New Haven: Yale Univ. Press.
— (1964a). Microbiological fractionation of sulphur isotopes. *J. gen. Microbiol*, **34**, 195–212.
— (1964b). Carbon isotope fractionation during metabolism of lactate by *Desulfovibrio desulfuricans*. *J. gen. Microbiol.*, **34**, 213–17.
Karrer, P. (1960). *Organic Chemistry*. Amsterdam: Elsevier.
Keith, S. M., Herbert, R. A. & Harfoot, C. G. (1982). Isolation of new types of sulphate-reducing bacteria from estuarine and marine sediments using chemostat enrichments. *J. appl. Bact.*, **53**, 29–33.

Kellog, W. W., Cadle, R. D., Allen, E. R., Lazrus, A. L. & Martell, E. A. (1972). The sulphur cycle. *Science*, **175**, 587–96.

Kelly, D. P. (1982). The sulphur bacteria: concluding remarks. *Phil. Trans. R. Soc. Lond. B.*, **298**, 601–2.

Kent, T. A., Moura, I., Moura, J. J. G., Lipscomb, J. D., Huynh, B. H., Le Gall, J., Xavier, A. V. & Münck, E. (1982). Conversion of [3Fe–3S] into [4Fe–4S] clusters in a *Desulfovibrio gigas* ferredoxin and isotopic labelling of iron-sulfur cluster subsites. *FEBS Lett*, **138**, 55–8.

Kerscher, L. & Oesterhelt, D. (1982). Pyruvate:ferredoxin oxidoreductase – new findings on an ancient enzyme. *Trends in Biochem. Sci.*, **7**, 371–4.

Kimata, M., Kadota, H. & Hata, Y. (1955a). Studies on the marine sulfate-reducing bacteria. IV. *Bull. Jap. Soc. scient. Fish.*, **21**, 229–34.

— (1955b). Studies on the marine sulfate-reducing bacteria. V. *Bull. Jap. Soc. scient. Fish.*, **21**, 235–9.

Kimata, M., Kadota, H., Hata, Y. & Miyoshi, H. (1958). The formation of sulfide by sulfate-reducing bacteria in the estuarine zone of the river receiving a large quantity of organic drainage. *Rec. oceanogr. Wks. Japan*, Special No. **2**, 187–99.

Kimata, M., Kadota, H., Hata, Y. & Tajima, T. (1955a, b, c). Studies on the marine sulfate-reducing bacteria. I, II and III. *Bull. Jap. Soc. scient. Fish.*, **21**, 102–8, 109–12, 113–18.

Kimura, K., Suzuki, A., Inokuchi, H. & Yagi, T. (1979). Hydrogenase activity in the dry state. Isotope exchange and reversible oxido-reduction of cytochrome c_3. *Biochim. biophys. Acta*, **567**, 96–105.

King, N. K. & Winfield, M. E. (1955). The assay of soluble hydrogenase. *Biochim. biophys. Acta*, **18**, 431–2.

Klein, R. M. & Cronquist, A. (1967). A consideration of the evolutionary and taxonomic significance of some biochemical, micromorphological, and physiological characters in the thallophytes. *Q. Rev. Biol.*, **42**, 105–296.

Kluyver, A. J. (1940). In *Martinus Willem Beijerinck, his life and work*, ed. van Iterson, G., den Dooren de Jong, L. E. & Kluyver, A. J., pp. 97–154. The Hague:Nihoff.

Kluyver, A. J. & Baars. J. K. (1932). Microbiology – on some physiological artefacts. *Proc. Roy. Soc. Amst.* **35**, 370–8.

Knivett, V. A. (1960). The microbiological production of vitamin B_{12} and sulphide from sewage. *Prog. ind. Microbiol.*, **2**, 29–45.

Knolles, A. S. (1952). Some preliminary experiments in the filling of waterlogged pits with refuse. *J. Roy. sanit. Inst.*, **72**, 55–63.

Kobayashi, K., Hasegawa, H., Takagi, M. & Ishimoto, M. (1982). Proton translocation associated with sulfite reduction in a sulfate-reducing bacterium, *Desulfovibrio vulgaris*. *FEBS Lett.*, **142**, 235–7.

Kobayashi, K. & Skyring, G. W. (1982). Ultrastructural and biochemical characterization of Miyazaki strains of *Desulfovibrio vulgaris*. *J. gen. appl. Microbiol.*, *Toyko*, **28**, 45–54.

Kobayashi, K., Tachibana, S. & Ishimoto, M. (1969). Intermediary formation of trithionate in sulfite reduction by a sulfate-reducing bacterium. *J. Biochem, Tokyo*, **65**, 155–7.

Kobayashi, K., Takahashi, E. & Ishimoto, M. (1972). Biochemical studies on sulfate-reducing bacteria. XI. Purification and some properties of sulfite reductase, desulfoviridin. *J. Biochem., Tokyo*, **72**, 879–87.

Kornberg, H. L. (1966). Anaplerotic sequences and their role in biochemistry. In *Essays in biochemistry*, ed. Campbell, P.N. & Greville, G. D., vol. 2, pp. 1–31, London: Academic Press.

Koyama, J., Tamiya, N., Ishimoto, M. & Nagai, H. (1954). (Title in Japanese). *Seikagaku zasshi*, **26**, 304.
Koyama, T. & Sugawara, K. (1953). Sulphur metabolism in bottom muds and related problems. *J. Earth Sci. Nagoya Univ.*, **1**, 24–34.
Krasna, A. I. & Rittenberg, D. (1954). The mechanism of action of the enzyme hydrogenase. *J. Amer. chem. Soc.*, **76**, 3015–20.
— (1955). The reduction of nitroprusside by hydrogen with *Proteus vulgaris*. *J. Amer. chem. Soc.*, **77**, 5295–6.
Krassowski, B., Sadurska, I. & Kowalik, R. (1966). Contamination of chrome-tanned leather by *Desulfovibrio*. *Acta Microbiol. Pol.*, **15**, 203–4.
Krieg, N. R. (ed.) (1983). *Bergey's Manual of Systematic Bacteriology*, vol. 1. Baltimore: The Williams and Wilkins Co.
Kristjansson, J. K., Schönheit, P. & Thauer, R. K. (1982). Different K_s values for hydrogen of methanogenic bacteria and sulfate reducing bacteria: an explanation for the apparent inhibition of methanogenesis by sulfate. *Arch. Microbiol.*, **131**, 278–82.
Krumbein, W. E. & Pochon, J. (1964). Ecologie bactérienne des pierres altérées des monuments. *Annls. Inst. Pasteur, Paris*, **107**, 724–32.
Kuenen, J. G. & Veldkamp, H. (1972). *Thiomicrospira pelophila* gen. n., sp. n., a new obligately chemolithotrophic colourless sulfur bacterium. *Antonie van Leeuwenhoek*, **38**, 241–56.
Kutznetsova, V. A., Li, A. D. & Tiforova, N. N. (1963). A determination of source of contamination of oil-bearing D_1 strata of Romashkino field by sulfate-reducing bacteria. *Mikrobiologiya*, **32**, 683–7 (T581–585).
Kutznetsova, V. A. & Pantskhava, E. S. (1962). Effect of freshening of stratal waters on development of halophilic sulfate-reducing bacteria. *Mikrobiologiya*, **31**, 129–34 (T103–106).
Laanbroek, H. J. & Pfennig, N. (1981). Oxidation of short-chain fatty acids by sulfate-reducing bacteria in fresh-water and in marine sediments. *Arch. Microbiol.*, **128**, 330–5.
Laanbroek, H. J. & Veldkamp, H. (1982). Microbial interactions in sediment communities. *Phil. Trans. R. Soc. Lond. B.*, **297**, 533–50.
La Rivière, J. W. M. (1955*a*). The production of surface active compounds by micro-organisms and its possible significance in oil recovery. I. Some general observations on the change of surface tension in microbial cultures. *Antonie van Leeuwenhoek*, **21**, 1–8.
— (1955*b*). The production of surface active compounds by micro-organisms and its possible significance in oil recovery. II. On the release of oil from oil-sand mixtures with the aid of sulphate-reducing bacteria. *Antonie van Leeuwenhoek*, **21**, 9–27.
Laube, V. M. & Martin, S. M. (1981). Conversion of cellulose to methane and carbon dioxide by triculture of *Acetovibrio cellulolyticus*, *Desulfovibrio* sp. and *Methanosarcina barkeri*. *Appl. env. Microbiol.*, **42**, 413–20.
Lawrance, W. A. (1948). Determination of the time of passage of pollution in the Androscoggin River and pond. *Sewage Wks. J.*, **20**, 881–96.
— (1950). The addition of sodium nitrate to the Androscoggin River. *Sewage & Ind. Wastes.*, **22**, 820–32.
Lee, J. P. & Peck, H. D. (1971). Purification of the enzyme reducing bisulfite to trithionate from *Desulfovibrio gigas* and its identification as desulfoviridin. *Biochem. biophys. Res. Commun.*, **45**, 583–9.

184 References

Lee, J. P., Yi, C. S., Le Gall, J. & Peck, H. D. (1973). Isolation of a new pigment desulforubidin, from *Desulfovibrio desulfuricans* (Norway strain) and its role in sulfite reduction. *J. Bact.*, **115**, 453–5.

Le Gall, J. & Bruschi-Heriaud, M. (1968). Purification and properties of a new cytochrome from *D. desulfuricans*. In *Structure and function of cytochromes*, ed. Okunuki, K., Kamen, M. D. & Sekusu, I. pp. 467–70. Baltimore: Univ. Park Press.

Le Gall, J., Bruschi-Heriaud, M. & DerVartanian, D. V. (1971a). Electron paramagnetic resonance and light absorption studies on c-type cytochromes of the anaerobic sulfate reducer *Desulfovibrio*. *Biochim. biophys. Acta*, **234**, 499–512.

Le Gall, J., DerVartanian, D. V. & Peck, H. D. (1979). Flavoproteins, iron proteins and hemoproteins as electron-transfer components of the sulfate reducing bacteria. *Current Topics in Bioenergetics*, **9** 237–65.

Le Gall, J., DerVartanian, D. V., Spilker, E., Lee, J.P. & Peck, H. D. (1971b). Evidence for the involvement of non-heme iron in the active site of hydrogenase from *Desulfovibrio vulgaris*. *Biochim. biophys. Acta*, **234**, 525–30.

Le Gall, J., Ljungdahl, P. O., Moura, I., Peck, H. D., Xavier, A. V., Moura, J. J. G., Teixera, M., Huynh, B. H. & DerVartanian, D. V. (1982). The presence of redox-sensitive nickel in the periplasmic hydrogenase from *Desulfovibrio gigas*. *Biochem. biophys. Res. Commun.*, **106**, 610–16.

Le Gall, J., Mazza, G. & Dragoni, N. (1965). Le cytochrome c_3 de *Desulfovibrio gigas*. *Biochim. biophys. Acta*, **56**, 385–7.

Le Gall, J. & Postgate, J. R. (1973). The physiology of sulphate-reducing bacteria. *Adv. microb. Physiol.*, **10**, 81–133.

Le Gall, J., Senez, J. C. & Pichinoty, F. (1959). Fixation de l'azote par les bactéries sulfato-réductrices: isolement et caraterisation de souches actives. *Annls. Inst. Pasteur, Paris*, **96**, 223–30.

Levin, R. E.., Ng. H., Nagel, C. W. & Vaughn, R. H. (1959). Desulfovibrios associated with hydrogen sulfide formation in olive brines. *Bact. Proc.*, P7.

Lewis, A. J. & Miller, J. D. A. (1975). Keto acid metabolism in Desulfovibrio. *J. gen. Microbiol.*, **90**, 286–92.

— (1977). The tricarboxylic acid pathway in Desulfovibrio. *Can. J. Microbiol*, **23**, 916–21.

Lewis, D. (1954). The reduction of sulphate in the rumen of the sheep. *Biochem. J.*, **56**, 391–9.

Lewis, K. (1966). Symposium on bioelectrochemistry in micro-organisms. IV. Biochemical fuel cells. *Bact. Rev.*, **30**, 101–13.

Liberthson, L. (1945). Bacterial deterioration of cutting oil emulsions. *Lubrication engng.*, **1** 103–6.

Lighthart, B. (1963). Sulfate-reducing bacteria in San Vincente Reservoir, San Diego County, California. *Limnol. Oceanogr.*, **8**, 349–51.

Lin, C. C. & Lin, K. C. (1970). Spoilage bacteria in canned foods. II. Sulfide spoilage bacteria in canned mushrooms and a versatile medium for the enumeration of *Clostridium nigrificans*. *Appl. Microbiol.*, **19**, 283–6.

Littlewood, D. & Postgate, J. R. (1956). Substrate inhibition of hydrogenase enhanced by sodium chloride. *Biochim. biophys. Acta*, **20**, 399–400.

— (1957a). Sodium chloride and the growth of *Desulphovibrio desulphuricans*. *J. gen. Microbiol.*, **17**, 378–89.

— (1957b). On the osmotic behaviour of *Desulphovibrio desulphuricans. J. gen. Microbiol*, **16**, 596–603.

Liu, C. L., DerVartanian, D. V. & Peck, H. D. (1979). On the redox properties of three bisulfite reductases from the sulfate-reducing bacteria. *Biochem. biophys. Res. Commun.*, **91**, 962–70.

Liu, Chi-Li & Peck, H. D. (1981). Comparative bioenergetics of sulfate reduction in *Desulfovibrio* and *Desulfotomaculum* spp. *J. Bact.*, **145**, 966–73.

Liu, M. C., DerVartanian, D. V. & Peck, H. D. (1980). On the nature of the oxidation–reduction properties of nitrite reductase from *Desulfovibrio desulfuricans*. *Biochem. biophys. Res. Commun.*, **96**, 278–85.

Liu, M. C. & Peck, H. D. (1981). The isolation of a hexaheme cytochrome from *Desulfovibrio desulfuricans* and its identification as a new type of nitrite reductase. *J. biol. Chem.*, **256**, 13159–64.

Lwoff, A. (1944). *L'evolution physiologique*. Paris: Masson.

McCready, R. G. L. & Krouse, H. R. (1980). Sulfur isotope fractionation by *Desulfovibrio vulgaris* during metabolism of $BaSO_4$. *Geomicrobiol. J.*, **2**, 55–62.

McCready, R. G. L., Gould, W. G. & Barendregt, R. W. (1983). Nitrogen isotope fractionation during the reduction of NO_2^- to NH_4^+ by *Desulfovibrio* sp. *Can. J. Microbiol.*, **29**, 231–4.

McDonald, C. C., Phillips, W. D. & Le Gall, J. (1974). Proton magnetic resonance studies of *Desulfovibrio* cytochromes c_3. *Biochemistry*, **13**, 1952–8.

McInerney, M. J. & Bryant, M. P. (1981). Anaerobic degradation of lactate by syntrophic associations of *Methanosarcina barkeri* and *Desulfovibrio* species and effect of H_2 on acetate degradation. *Appl. env. Microbiol.*, **41**, 346–54.

McInerney, M. J., Bryant, M. P., Hespell, R. B. & Costerton, J. W. (1981). *Syntrophomonas wolfii* gen. nov., sp. nov., an anaerobic syntrophic fatty acid-oxidizing bacterium. *Appl. env. Microbiol.*, **41**, 1029–39.

McInerney, M. J., Bryant, M. P. & Pfennig, N. (1979). Anaerobic bacterium that degrades fatty acids in syntrophic association with methanogens. *Arch. Microbiol*, **122**, 129–35.

McInerney, M. J., Mackie, R. I. & Bryant, M. P. (1981). Syntrophic association of a butyrate-decomposing bacterium and *Methanosarcina* enriched from bovine rumen fluid. *Appl. env. Microbiol*, **41**, 826–8.

McKinney, R. E. & Conway, R. A. (1957). Chemical oxygen in biological waste treatment. *Sewage & Ind. Wastes*, **29**, 1097–1106.

Magee, E. L., Ensley, B. D. & Barton, L. L. (1978). An assessment of growth yields and energy coupling in *Desulfovibrio*. *Arch. Microbiol.*, **117**, 21–6.

Mager, J., Kuczynski, M., Schatzberg, G. & Avi-dor, Y. (1956). Turbidity changes in bacterial suspensions in relation to osmotic pressure. *J. gen. Microbiol.*, **14**, 69–75.

Mara, D. D. & Williams, D. J. A. (1970). The evaluation of media used to enumerate sulphate-reducing bacteria. *J. appl. Bact.*, **33**, 543–52.

Maroc, J., Azerad, R., Kamen, M. D. & Le Gall, J. (1970). Menaquinone (MK-6) in the sulfate-reducing obligate anaerobe, *Desulfovibrio*. *Biochim. biophys. Acta*, **197**, 87–9.

Martin, S. M., Glick, B. R. & Martin, W. G. (1980). Factors affecting the production of hydrogenase by *Desulfovibrio desulfuricans*. *Can. J. Microbiol.*, **26**, 1209–13.

Matheron, R. & Baulaigue, R. (1976). Bactéries fermentatives, sulfato-réductrices et phototrophes sulfureuses en cultures mixtes. *Arch. Microbiol*, **109**, 319–20.

Mechalas, B. J. & Rittenberg, S. C. (1960). Energy coupling in *Desulfovibrio desulfuricans*. *J. Bact.*, **80**, 501–7.

Mel, H. C. (1954). Chemical thermodynamics of aqueous thiosulfate and bromate ions. *Chem. Abstr.*, **48**, 6228.

Meyer, T. E., Bartsch, R. G. & Kamen, M. D. (1971). A class of electron transfer heme proteins found in both photosynthetic and sulfate-reducing bacteria. *Biochim. biophys. Acta*, **245**, 453–64.

Miller, J. D. A. (1971). *Microbial aspects of metallurgy*. Aylesbury: Medical & Technical Publn Co.

— (1981). Metals. In *Economic microbiology, vol. 6: Microbial biodeterioration*, ed. Rose, A. H., pp. 149–202. London: Academic Press.

Miller, J. D. A., Neumann, P. M., Elford, L. & Wakerley, D. S. (1970). Malate dismutation by Desulfovibrio. *Arch. Microbiol.*, **71**, 214–19.

Miller, J. D. A. & Saleh, A. M. (1964). A sulphate-reducing bacterium containing cytochrome c_3 but lacking desulfoviridin. *J. gen. Microbiol.*, **37**, 419–23.

Miller, J. D. A. & Wakerley, D. S. (1966). Growth of sulphate-reducing bacteria by fumarate dismutation. *J. gen. Microbiol.*, **43**, 101–7.

Miller, L. P. (1950). Formation of metal sulfides through the activities of sulfate-reducing bacteria. *Contrib. Boyce Thompson Inst.*, **16**, 85–9.

Millet, J. (1954). Dégradation anaérobie du pyruvate par un extrait enzymatique de *Desulfovibrio desulfuricans*. *C. r. hebd. Séanc. Acad. Sci., Paris*, **238**, 408–11.

— (1955). Le sulfite comme intermediaire dans la réduction du sulfate par *Desulfovibrio desulfuricans*. *C. r. hebd. Séanc. Acad. Sci, Paris*, **240**, 253–5.

Monster, J., Appel, P. W. U., Thode, H. G., Schidlowski, M., Carmichael, C. M. & Bridgwater, D. (1979). Sulfur isotope studies in early Archaean sediments from Isua, West Greenland: implications for the antiquity of bacterial sulfate reduction. *Geochim. Cosmochim. Acta*, **43**, 405–13.

Moore, W. E. C., Johnson, J. L. & Holderman, L. V. (1976). Emendations of *Bacteriodaceae* and *Butyrivibrio* and descriptions of *Desulfomonas* gen. nov. and ten new species in the genera *Desulfomonas, Butyrivibrio, Eubacterium, Clostridium* and *Ruminococcus*. *Intern. J syst. Bact.*, **26**, 238–52.

Morgan, G. B. & Lackey, J. B. (1965). Ecology of a sulfuretum in a semi-tropical environment. *Z. allg. Mikrobiol*, **5**, 237–48.

Morris, J. G. (1975). The physiology of obligate anaerobiosis. *Adv. microb. Physiol.*, **12**, 169–246; see footnote to p. 245.

Morton, M. W., Matschiner, J. T. & Peck, H. D. (1970). Menaquinone-6 in the stric anaerobes *Desulfovibrio vulgaris* and *Desulfovibrio gigas*. *Biochem. biophys. Res. Commun.*, **8**, 197–204.

Mountfort, D. O. & Asher, R. M. (1979). Effect of inorganic sulfide in growth and metabolism of *Methanosarcina barkeri* strain DM. *Appl. env. Microbiol.*, **37**, 670–75.

Moura, I., Bruschi, M., Le Gall, J., Moura, J. J. G. & Xavier, A. V. (1977b). Isolation an characterization of desulforedoxin, a new type of non-heme iron protein from *Desulfovibrio gigas*. *Biochem. biophys. Res. Commun.*, **75**, 1037–44.

Moura, I., Huynh, B. H., Hausinger, R. P., Le Gall, J., Xavier, A. V. & Münck, E. (1980 Mössbauer and e.p.r. studies of desulforedoxin from *Desulfovibrio gigas*. *J. biol. Chem* **255**, 2493–8.

Moura, I., Moura, J. J. G., Bruschi, M. & Le Gall, J. (1980). Flavodoxin and rubredoxin from *Desulphovibrio salexigens*. *Biochim. biophys. Acta.*, **591**, 1–8.

Moura, I., Moura, J. J. G., Santos, M. H., Xavier, A. V. & Le Gall, J. (1979). Redox studies on rubredoxins from sulphate and sulphur reducing bacteria. *FEBS Lett.*, **107**, 419–21.

Moura, J. J. G., Moura, I., Bruschi, M., Le Gall, J. & Xavier, A. V. (1980). A cobalt containing protein isolated from *Desulfovibrio gigas*, a sulfate reducer. *Biochem. biophys. Res. Commun.*, **92**, 962–70.

Moura, J. J. G., Moura, I., Huynh, B. H., Krüger, H. J., Teixeira, M., DuVarney, R. C., DerVartanian, D. V., Xavier, A. V., Peck, Jr., H. D. & Le Gall, J. (1982). Unambiguous identification of the nickel e.p.r. signal in ^{61}Ni-enriched *Desulfovibrio gigas* hydrogenase. *Biochem. biophys. Res. Commun.*, **108**, 1388–93 *Erratum*, **109**, 1073.

Moura, J. J. G., Xavier, A. V., Bruschi, M., Le Gall, J., Hall, D. O. & Cammack, R. (1976). A molybdenum-containing iron-sulphur protein from *Desulfovibrio gigas*. *Biochem. biophys. Res. Commun.*, **72**, 782–9.

Moura, J. J. G., Xavier, A. V., Bruschi, M. & Le Gall, J. (1977*a*). NMR characterization of three forms of ferredoxin from *Desulphovibrio gigas*, a sulphate reducer. *Biochim. biophys. Acta*, **459**, 278–89.

Moura, J. J. G., Xavier, A. V., Cookson, D. J., Moore, G. R., Williams, R. J. P., Bruschi, M. & Le Gall, J. (1977*c*). Redox states of cytochrome c$_3$ in the absence and presence of ferredoxin. *FEBS Lett.*, **81**, 275–80.

Moura, J. J. G., Xavier, A. V., Hatchikian, E. C. & Le Gall, J. (1978). Structural control of the redox potentials and of the physiological activity by oligomerization of ferredoxin. *FEBS Lett.*, **89**, 177–9.

Murphy, M. J. & Siegel, L. M. (1973). Siroheme and sirohydrochlorin. *J. biol. Chem.*, **248**, 6911–19.

Murphy, M. J., Siegel, L. M., Kamin, H., DerVartanian, D.V., Lee, J. P., Le Gall, J. & Peck, H. D. (1973). An iron tetrahydroporphyrin prosthetic group common to both assimilatory and dissimilatory sulfite reductases. *Biochem. biophys. Res. Commun.* **54**, 82–8.

Nakatsukasa, W. & Akagi, J. M. (1969). Thiosulfate reductase isolated from *Desulfotomaculum nigrificans*. *J. Bact.*, **98**, 429–33.

Nakos, G. & Mortenson, L. E. (1971). Structural properties of hydrogenase from *Clostridium pasteurianum* W5. *Biochemistry*, **10**, 2442–9.

Nazina, T. N. & Rozanova, E. P. (1978). Thermophilic sulfate-reducing bacteria from oil strata. *Mikrobiologiya*, **47**, 142–8 (T113–18).

Nazina, T. N., Rozanova, E. P. & Kalininskaya, T. A. (1979). Fixation of molecular nitrogen by sulfate-reducing bacteria from oil strata. *Mikrobiologiya*, **48**, 133–6 (T102–4).

Nedwell, D. B. & Abram, J. W. (1978). Bacterial sulfate reduction in relation to sulfur geochemistry in two contrasting areas of a salt marsh sediment. *Est. Coast Mar. Sci.*, **6**, 341–51.

Nedwell, D. B. & Abram, J. W. (1979). Relative influence of temperature and electron donor and electron acceptor concentrations on bacterial sulfate reduction in salt marsh sediment. *Microb. Ecol.*, **5**, 67–72.

Nedwell, D. B. & Azni bin Abdul Aziz, S. (1980). Heterotrophic nitrogen fixation in an intertidal saltmarsh sediment. *Estuarine and Coastal Marine Science*, **10**, 699–702.

Nedwell, D. B. & Banat, I. M. (1982). Hydrogen as an electron donor for sulfate-reducing bacteria in slurries of salt marsh sediment. *Microb. Ecol.*, **7**, 305–13.

Neuberg, C. & Mandl, I. (1948). An unknown effect of amino acids. *Archs Biochem.*, **19**, 149–61.

Nielands, J. B. (1973). Microbial iron transport compounds (siderochromes). In *Inorganic biochemistry*, ed Eichorn, G. L., pp. 167–202. New York: Elsevier.

Noordam-Goedewagen, M. A., Manten, A. & Muller, F. M. (1949). The influence of sulphide on the methane fermentation of sodium and calcium acetates. *Antonie van Leeuwenhoek*, **15**, 65–85.

Novelli, G. D. & ZoBell, C. E. (1944). Assimilation of petroleum hydrocarbons by sulfate-reducing bacteria. *J. Bact.*, **47**, 447–8.

O'Brien, R. W. & Stern, J. R. (1969). Reversal of the stereospecificity of the citrate synthase of *Clostridium kluyveri* by *p*-chloromercuribenzoate. *Biochem. biophys. Res. Commun.*, **34**, 271–6.

Ochynski, F. W. & Postgate, J. R. (1963). Some biological differences between fresh water and salt water strains of sulphate-reducing bacteria. In *Marine microbiology*, ed. Oppenheimer, C. H., pp. 426–41. Springfield, Ill., : C. C. Thomas.

Odom, J. M. & Peck, H. D. (1981a). Localization of dehydrogenases, reductases, and electron transfer components in the sulfate-reducing bacterium *Desulfovibrio gigas*. *J. Bact.*, **147**, 161–9.

— (1981b). Hydrogen cycling as a general mechanism for energy coupling in the sulfate-reducing bacteria, *Desulfovibrio* sp. *FEMS Microbiol. Lett.*, **12**, 47–50.

Ogata, M., Arihara, K. & Yagi, T. (1981). D-lactate dehydrogenase of *Desulfovibrio vulgaris*. *J. Biochem., Tokyo*, **89**, 1423–31.

Olson, G. J., Dockins, W. S., McFeters, G. A. & Iverson, W. P. (1981). Sulfate-reducing and methanogenic bacteria from deep aquifers in Montana. *Geomicrobiol. J.*, **2**, 327–40.

Oppenheimer, C. H. (1965). Bacterial production of hydrocarbon-like materials. *Z. allg. Mikrobiol.*, **5** 284–307.

Oremland, R. S. & Taylor, B. F. (1978). Sulfate reduction and methanogenesis in marine sediments. *Geochim. Cosmochim. Acta*, **42**, 209–14.

Pace, B. & Campbell, L. L. (1971). Homology of ribosomal ribonucleic acid of Desulfovibrio species with *Desulfovibrio vulgaris*. *J. Bact.*, **106**, 717–19.

Pankhurst, E. S. (1967). A simple culture tube for anaerobic bacteria. *Lab. Pract.*, **16** 58–9.

— (1968a). Significance of sulphate-reducing bacteria to the gas industry: a review. *J. appl. Bact.*, **31**, 179–93.

— (1968b). Bacteriological aspects of the storage of gas underground. *J. appl. Bact.*, **31** 311–22.

— (1971). The isolation and enumeration of sulphate-reducing bacteria. In *Isolation c anaerobes*, ed. Shapton, D. A. & Board, R. G., pp. 223–40. London: Academic Press.

Payne, W. J. & Grant, M. A. (1982). Influence of acetylene on growth of sulfate-respirin, bacteria. *Appl. env. Microbiol.*, **43**, 727–30.

Peck, H. D. (1959). The ATP-dependent reduction of sulfate with hydrogen in extracts c *Desulfovibrio desulfuricans*. *Proc. natn. Acad. Sci. USA*, **45**, 701–8.

— (1960). Evidence for oxidative phosphorylation during the reduction of sulfate wit hydrogen by *Desulfovibrio desulfuricans*. *J. biol. Chem*, **235**, 2734–8.

— (1962). Symposium on metabolism of inorganic compounds. V. Comparative metabolis of inorganic sulfur compounds in microorganisms. *Bact. Rev.*, **26**, 67–94.

References

— (1966). Phosphorylation coupled with electron transfer in extracts of the sulfate reducing bacterium, *Desulfovibrio gigas*. *Biochem. biophys. Res. Commun.*, **22**, 112–18.

— (1966–7). Some evolutionary aspects of inorganic sulfur metabolism. *Lectures on Theoretical and Applied Aspects of Modern Microbiology*. Univ. Park: Univ. Maryland Press.

— (1968). Energy-coupling mechanisms in chemolithotrophic bacteria. *Ann. Rev. Microbiol*, **22**, 489–518.

— (1974). The evolutionary significance of inorganic sulphur metabolism. In *Evolution in the microbial world*, ed Carlile, M. J. & Skehel, J. J., 24th Symp. Soc. gen. Microbiol., pp. 241–62. Cambridge University Press.

Peck, H. D. & Le Gall, J. (1982). Biochemistry of dissimilatory sulphate reduction. *Phil. Trans. R. Soc. Lond. B..*, **298**, 443–66.

Pfennig, N. (1978). Syntrophic associations and consortia with phototrophic bacteria. *Abs. 12. Int. Congr. Microbiol. (Munich)* Section 11.6, p. 16.

Pfennig, N. & Biebl, H. (1976). *Desulfuromonas acetoxidans* gen. nov. and sp. nov., a new anaerobic, sulfur-reducing, acetate-oxidizing bacterium. *Arch. Microbiol.*, **110**, 3–12.

Pfennig, N. & Biebl, H. (1981). The dissimilatory sulfur-reducing bacteria. In *The Prokaryotes. A handbook on habitats, isolation, and identification of bacteria*, ed. Starr, M. P., Stolp, H., Trüper, H.G., Balows, A. & Schlegel, H. G., vol. I. pp. 941–7. Berlin: Springer-Verlag.

Pfennig, N. & Widdel, F. (1982). The bacteria of the sulphur cycle. *Phil. Trans. R. Soc. Lond. B*, **298**, 433–41.

Pfennig, N., Widdel, F. & Trüper, H. G. (1981). The dissimilatory sulfate-reducing bacteria. In *The Prokaryotes. A handbook on habitats, isolation, and identification of bacteria*, ed. Starr, M. P., Stolp, H., Trüper, H. G., Balows, A. & Schlegel, H. G., vol. I, pp. 926–40. Berlin: Springer-Verlag.

Pichinoty, F. & Senez, J. C. (1958). Couplage entre la dèshydrogenation du pyruvate et la reduction du nitrite chez les bactéries sulfatoréductrices. Nature du transporteur d'hydrogene. *C. r. hebd. Acad. Séanc. Acad. Sci., Paris*, **247**, 361–4.

Pierrot, M., Haser, R., Frey, M., Bruschi, M., Le Gall, J., Sieker, L. & Jensen, L. H. (1976). Some comparisons between two crystallized anaerobic bacterial rubredoxins from *Desulfovibrio gigas* and *D. vulgaris*. *J. mol. Biol.*, **107**, 179–82.

Pipes, W. O. (1960). Sludge digestion by sulphate-reducing bacteria. *Proc. 15th Indust. Waste Conf., Purdue Univ. Engng, Extn. Ser.*, **106**, 308–19.

Plessis, A. & Gatellier, M. C. (1965). Le role des associations bactériennes dans le developpement et la resistance des reducteurs de sulfate. *C. r. congr. Int. Corros. Mar., Cannes*, 377–80.

Pochon, J. (1955). Technique de preparation des suspensions – dilutions de terre pour analyse microbiologique. *Annls. Inst. Pasteur, Paris*, **89**, 464–5.

Pochon, J. & Chalvignac, M. A. (1952). Sur l'instabilité des caractères d'une souche de *Sporovibrio*. *Annls. Inst. Pasteur, Paris*, **82**, 399–403.

Pochon, J., Coppier, O. & Tchan, Y. T. (1951). Role des bactéries dans certaines alterations des pierres des monuments. *Chim. Ind.*, **65**, 496–500.

Pochon, J. & de Barjac, H. (1954). Une espece nouvelle de *Sporovibrio*: Sp. *ferroxydans* (n. sp.). *C. r. hebd. Séanc. Acad. Sci., Paris*, **238**, 627–8.

Postgate, J. R. (1949). Competitive inhibition of sulphate reduction by selenate. *Nature, Lond.*, **164**, 670–1.

— (1951a). On the nutrition of *Desulfovibrio desulfuricans*. *J. gen. Microbiol.*, **5**, 714–24.
— (1951b). The reduction of sulphur compounds by *Desulphovibrio desulphuricans*. *J. gen. Microbiol.*, **5**, 725–38.
— (1952a). Growth of sulphate-reducing bacteria in sulphate-free media. *Research, Lond.*, **5** 189–190.
— (1952b). Competitive and non-competitive inhibitors of bacterial sulphate reduction. *J. gen. Microbiol.*, **6**, 128–42.
— (1953a). On the nutrition of *Desulphovibrio desulphuricans*: a correction. *J. gen. Microbiol.*, **9**, 440–4.
— (1953b). Discussion. In *Adaptation in micro-organisms*, ed. Gale, E. F. & Davies, R., p. 324. Cambridge Univ. Press.
— (1954). Presence of cytochrome in an obligate anaerobe. *Biochem. J.*., **56**, xi.
— (1956a). Cytochrome c_3 and desulphoviridin; pigments of the anaerobe *Desulphovibrio desulphuricans*. *J. gen. Microbiol.*, **14**, 545–72.
— (1956b). Iron and the synthesis of cytochrome c_3. *J. gen. Microbiol.*, **15**, 186–93.
— (1956c). Sulphate-reducing bacteria which are deficient in cytochrome. *J. gen. Microbiol.*, **15**, viii.
— (1959a). Sulphate reduction by bacteria. *A. Rev. Microbiol.*, **13**, 505–20.
— (1959b). A diagnostic reaction for *Desulphovibrio desulphuricans*. *Nature, Lond.*, **183**, 481–2.
— (1960a). The economic activities of sulphate-reducing bacteria. *Prog. ind. Microbiol.*, **2**, 48–69.
— (1960b). On the autotrophy of *Desulphovibrio desulphuricans*. *Z. allg. Mikrobiol.*, **1** 53–56.
— (1961). Cytochrome c_3. In *Haematin enzymes*, ed. Falk, J., Lamborg, R. & Morton, R. K., pp. 407–414. London: Pergamon Press.
— (1963a). Sulfate-free growth of *Clostridium nigrificans*. *J. Bact.*, **85**, 1450–1.
— (1963b). A strain of *Desulfovibrio* able to use oxamate. *Arch. Mikrobiol.*, **46**, 287–95.
— (1963c). The microbiology of corrosion. The *Corrosion handbook*, vol. 1, ed. Shreir, L. L., pp. 2.51–2.64. London: Newnes.
— (1965a). Recent advances in the study of the sulfate-reducing bacteria. *Bact. Rev.*, **29**, 425–41.
— (1965b). Enrichment and isolation of sulphate-reducing bacteria. In *Anreichungskultur und Mutantenauslese*, ed Schlegel, H. G., & Kroger, E. pp. 190–7. Stuttgart: Fischer.
— (1968). Fringe biochemistry among microbes. *Proc. R. Soc. B*, **171**, 67–76.
— (1969a). Viable counts and viability. In *Methods in Microbiology*, vol. 1, ed. Norris, J. R. & Ribbons, D. W., pp. 611–628. London: Academic Press.
— (1969b). Methane as a minor product of pyruvate metabolism by sulphate-reducing and other bacteria. *J. gen. Microbiol.*, **57**, 293–302.
— (1970a). Nitrogen fixation by sporulating sulphate-reducing bacteria including rumen strains. *J. gen. Microbiol.*, **63**, 137–9.
— (1970b). Carbon monoxide as a basis for primitive life on other planets: a comment *Nature, Lond.*, **226**, 984.
— (1974). Evolution within nitrogen-fixing systems. In *Evolution in the Microbial World*, ed Carlisle, M. & Skehel, J. C.U.P. *Symp. Soc. gen. Microbiol.*, pp. 263–92.
— (1982a). Economic importance of sulphur bacteria. *Phil. Trans. R. Soc. Lond. B.*, **298** 583–600.
— (1982b). *The fundamentals of nitrogen fixation*. Cambridge Univ. Press.

Postgate, J. R. & Campbell, L. L. (1966). Classification of *Desulfovibrio* species, the nonsporulating sulfate-reducing bacteria. *Bact. Rev.*, **30**, 732–8.

Postgate, J. R. & Hunter, J. R. (1962). The survival of starved bacteria. *J. gen. Microbiol.*, **29**, 233–63; errata, *J. gen. Microbiol.*, (1964). **34**, 473.

Postgate, J. R. & Kelly, D. P. (eds.) (1982). The sulphur bacteria. *Phil. Trans. R. Soc. Lond. B.*, **298**, 429–602. London: Royal Society.

Probst, I., Bruschi, M., Pfennig, N. & Le Gall, J. (1977). Cytochrome c-551.5 (c_7) from *Desulfuromonas acetoxidans*. *Biochim. biophys. Acta*, **460**, 58–64.

Purkiss, B. E. (1971). Corrosion in industrial situations by mixed microbial floras. In *Microbial aspects of metallurgy*, ed. Miller, J. D. A., pp. 107–28. Med. Tech. Publn Co.: Aylesbury.

Rao, K. K. & Cammack, R. (1981). The evolution of ferredoxin and superoxide dismutase in micro-organisms. In *Molecular and cellular aspects of microbial evolution*, ed. Carlile, M. J., Collins, J. F., & Moseley, B. E. B. *Symp. Soc. gen. Microbiol.*, **32**, 175–213.

Riederer-Henderson, M. A. & Peck, H. D. (1970). Formic dehydrogenase of *Desulfovibrio gigas*. *Bact. Proc.*, P70.

Riederer-Henderson, M. A. & Wilson, P. W. (1970). Nitrogen fixation by sulphate-reducing bacteria. *J. gen. Microbiol.*, **61**, 27–32.

Rittenberg, S. C. (1941). *Studies on marine sulfate-reducing bacteria*. Ph. D. dissertation: Univ. of California.

— (1969). The roles of exogenous organic matter in the physiology of chemolithotrophic bacteria. *Adv. microb. Physiol.*, **3**, 159–96.

Rogers, T. H. (1940). The inhibition of sulphate-reducing bacteria by dyestuffs. *J. Soc. Chem. Ind.*, **59**, 34–9.

— (1945). The inhibition of sulphate-reducing bacteria by dyestuffs. Part II. Practical applications in cable storage tanks and gas holders. *J. Soc. chem. Ind., Lond.*, **64**, 292–5.

Römer, R. & Schwartz, W. (1965). Geomikrobiologische untersuchungen V. Verwertung von sulfatmineralien und schwermetall-toleranz bei Desulfurizierern. *Z. allg. Mikrobiol*, **5**, 122–35.

Rosenfeld, W. D. (1948). Fatty acid transformations by anaerobic bacteria. *Archs. Biochem.*, **16**, 263–73.

Rossini, F. D., Wagman, D. D., Evans, W. H., Lavine, S. & Jaffe, I. (1952). *Selected values of chemical thermodynamic properties*. USA: Circular US Bureau of Standards No. 500.

Rozanova, E. P. & Khudyakova, A. I. (1974). A new non-spore-forming thermophilic sulfate-reducing organism, *Desulfovibrio thermophilus* nov. sp. *Mikrobiologiya*, **43**, 1069–75 (T908–12).

Rozanova, E. P. & Nazina, T. N. (1976). Mesophilic rod-like non-spore-forming bacterium reducing sulphates. *Mikrobiologiya*, **45**, 825–30 (T711–16).

Russell, P. (1961). Microbiological studies in relation to moist groundwood pulp. *Chem Ind.*, 642–9.

Rutten, M. (1972). *The origin of life by natural causes*. Amsterdam: Elsevier.

Sabina, L. R. & Pivnick, H. (1956). Oxidation of soluble oil emulsions and emulsifiers by *Pseudomonas oleovorans* and *Pseudomonas formicans*. *Appl. Microbiol.*, **4** 171–3.

Sadana, J. C. (1954). Pyruvate oxidation in *Desulphovibrio desulphuricans*. *J. Bact.*, **67**, 547–53.

Sadana, J. C. & Jagganathan, V. (1954). Purification of hydrogenase from *Desulfovibrio desulfuricans*. *Biochim. biophys. Acta*, **14**, 287–8.

Sadana, J. C. & Morey, A. V. (1961). Purification and properties of the hydrogenase of *Desulfovibrio desulfuricans*. *Biochim. biophys. Acta*, **50**, 153–63.

Saleh, A., Macpherson, R. & Miller, J. D. A. (1964). The effect of inhibitors on sulphate-reducing bacteria: a compilation. *J. appl. Bact.*, **27**, 281–93.

Saslawsky, A. S. & Chait, S. S. (1929). The influence of the concentration of sodium chloride on several biochemical processes in the liman. *Zentbl. Bakt. ParasitKde* (abt. 2), **77**, 18–21.

Schidlowski, M. (1980). The atmosphere. In *The handbook of environmental chemistry* 1 (A), ed. Hutzinger, O., pp. 1–16. Berlin, Heidelberg, New York: Springer.

Schoberth, S. (1973). A new strain of *Desulfovibrio gigas* isolated. *Arch. Mikrobiol.*, **92**, 365–8.

Schönheit, P., Kristjansson, J. K. & Thauer, R. K. (1982). Kinetic mechanism for the ability of sulfate reducers to out-compete methanogens for acetate. *Arch. Mikrobiol.*, **132**, 285–8.

Schwartz, R. M. & Dayhoff, M. O. (1978). Origins of prokaryotes, eukaryotes, mitochondria, and chloroplasts. *Science*, **199**, 395–403.

Sefer, M. & Calinescu, I. (1969). Bacterii sulfatreducatoare (genus Desulfovibrio), izolate din caria dentara, la om. *Microbiologia, Parazitologia, Epidemiologia (Hungary)*, **14**, 231–5.

Sefer, M. & Pozsgi, N. (1968). Etude serologique des bactéries sulfatoréductrices (genre *Desulfovibrio*) isolées en Roumanie. *Arch. Roum. exp. Microbiol.*, **27**, 867–73.

Seki, Y. & Ishimoto, M. (1979). Catalytic activity of the chromophore of desulfoviridin, sirohydrochlorin, in sulfite reduction in the presence of iron. *J. Biochem., Tokyo*, **86**, 273–6.

Seki, Y., Kobayashi, K. & Ishimoto, M. (1979). Biochemical studies on sulfate-reducing bacteria. XV. Separation and comparison of two forms of desulfoviridin. *J. Biochem., Tokyo*, **85**, 705–11.

Sekiguchi, T. & Nosoh, Y. (1973). Pyruvate-supported acetylene and sulfate reduction by cell-free extracts of *Desulfovibrio desulfuricans*. *Biochem. biophys. Res. Commun.*, **51**, 331–5.

Selwyn, S. C. & Postgate, J. R. (1959). A search for the *rubentschikii* group of *Desulphovibrio*. *Antonie van Leeuwenhoek*, **25**, 465–72.

Senez, J. C. (1952). Metabolisme des acides aminés et des amides par les bactéries sulfatoréductrices. *Annls. Inst. Pasteur, Paris*, **83**, 786–91.

— (1953). Sur l'activité et la croissance des bactéries anaérobies sulfato-réductrices en cultures semi-autotrophes. *Annls. Inst. Pasteur., Paris*, **84**, 595–605.

— (1954). Fermentation de l'acide pyruvique et des acides dicarboxyliques par les bactéries anaérobies sulfato-réductrices. *Bull. Soc. Chim. biol.*, **36**, 541–52.

— (1962). Some considerations of the energetics of bacterial growth. *Bact. Rev.*, **26**, 95–107.

Senez, J. C. & Cattaneo-Lacombe, J. (1956). Transformation de l'acide α-aspartique en α-alanine par des extraits de *Desulfovibrio desulfuricans*. *C. r. hebd. Séanc. Acad. Sci., Paris*, **242**, 941–3.

Senez, J. C., Geoffrey, C. & Pichinoty, F. (1956). Role des bactéries sulfato-réductrices dans la pollution des gasometres. *Gaz. de France Publn*, M-102.

Senez, J. C. & Leroux-Gilleron, J. (1954). Note preliminaire sur la degradation anaérobie de la cysteine et de la cystine par les bactéries sulfato-réductrices. *Bull. Soc. Chim. biol.*, **36**, 553–9.

Senez, J. C. & Pichinoty, F. (1958a). Réduction de l'hydroxylamine liée a l'activaté de

l'hydrogenase de *Desulfovibrio desulfuricans*: I. Activité des cellules et des extraits. *Biochim. biophys. Acta*, **27**, 569–80.
— (1958b). Réduction de l'hydroxylamine liée a l'activité de l'hydrogenase de *Desulfovibrio desulfuricans*: II. Nature du systeme enzymatique et du transporteur de electrons intervenant dans la réaction. *Biochim. biophys. Acta*, **28**, 355–69.
— (1958c). Reduction of nitrite by molecular hydrogen by *Desulfovibrio desulfuricans* and other bacteria. *Bull. soc. chim. biol.*, **40**, 2099–117.
Senez, J. C., Pichinoty, F. & Konavaltchikoff-Mazoyer, M. (1956). Réduction des nitrites et de l'hydroxylamine par les suspensions et les extraits de *Desulfovibrio desulfuricans*. *C. r. hebd. Séanc. Acad. Sci., Paris*, **242**, 570–3.
Shinkai, W., Hase, T., Yagi, T. & Matsubara, H. (1980). Amino acid sequence of cytochrome c_3 from *Desulfovibrio vularis*, Miyazaki. *J. Biochem., Tokyo*, **87**, 1747–56.
Siebenthal, C. E. (1915). Origin of the zinc and lead deposits of the Joplin region. *US Geol. Survey Bull*, **606**, 283.
Siegel, L. M. (1978). Structure and function of siroheme and siroheme enzymes. In *Mechanisms of oxidizing enzymes*, ed. Singer, T. P. & Ondarza, R. N., pp. 201–14. New York: Elsevier.
Sieker, L. C., Adman, E. & Jensen, L. H. (1971). Structure of the FeS complex in a bacterial ferredoxin. *Nature, Lond*, **235**, 40–2.
Sieker, L. C., Jensen, L. H., Bruschi, M., Le Gall, J., Moura, I. & Xavier, A. V. (1980). Desulforedoxin: preliminary X-ray diffraction study of a new iron-containing protein. *J. molec. Biol.*, **144**, 593–4.
Silverman, M. P. & Ehrlich, H. L. (1964). Microbial formation and degradation of minerals. *Adv. appl. Microbiol.*, **6**, 153–98.
Singleton, R., Campbell, L. L. & Hawkridge, F. M. (1979). Cytochrome c_3 from the sulfate-reducing anaerobe *Desulfovibrio africanus*: purification and properties. *J. Bact.*, **140**, 893–901; erratum, *J. Bact.*, (1981), **145**, 1212.
Sisler, F. D. (1961). Electrical energy from biochemical fuel cells. *New Scientist*, **12**, 110–11.
Sisler, F. D. & Senftle, F. E. (1963). Possible influence of the Earth's magnetic field on geomicrobiological processes in the hydrosphere. In *Marine microbiology*, ed. Oppenheimer, C. H., pp. 159–71. Springfield, Ill.: Thomas.
Sisler, F. D., Senftle, F. E. & Skinner, J. (1977). Electrobiochemical neutralization of acid mine water. *J. Water Pollution Control Fed.*, **1**, 369–74.
Sisler, F. D. & ZoBell, C. E. (1951a). Hydrogen utilization by some marine sulfate-reducing bacteria. *J. Bact.*, **62**, 117–27.
— (1951b). Nitrogen fixation by sulfate-reducing bacteria indicated by nitrogen/argon ratios. *Science*, **113**, 511–12.
Skyring, G. W. & Donnelly, T. H. (1982). Precambrian sulfur isotopes and a possible role for sulfite in the evolution of biological sulfate reduction. *Precambrian Res.*, **17**, 41–61.
Skyring, G. W. & Jones, H. E. (1972). Guanine plus cytosine contents of the deoxyribonucleic acids of some sulfate-reducing bacteria: a reassessment. *J. Bact.*, **109**, 1298–1300.
Skyring, G. W. & Jones, H. E. (1976). Variations in the spectrum of desulfoviridin from *Desulfovibrio gigas*. *Aust. J. biol. Sci.*, **29**, 291–9.
— (1977). Dithionite reduction in the presence of a tetrapyrrole-containing fraction from the desulfoviridin of *Desulfovibrio gigas*. *Aust. J. biol. Sci.* **30**, 21–31.
Skyring, G. W., Jones, H. E. & Goodchild, D. (1977). The taxonomy of some new isolates of dissimilatory sulfate-reducing bacteria. *Can. J. Microbiol* **23**, 1415–25.

Skyring, G. W., Oshrain, R. L. & Wiebe, W. J. (1979). Sulfate reduction rates in Georgia marshland soils. *Geomicrobiol. J.*, **1**, 389–96.

Skyring, G. W. & Trudinger, P. A. (1972). A method for the electrophoretic characterization of sulfite reductases in crude preparations from sulfate-reducing bacteria using polyacrylamide gels. *Can. J. Biochem.*, **50**, 1145–8.

— (1973). A comparison of the electrophoretic properties of the ATP-sulfurylases, APS-reductases, and sulfite reductases from cultures of dissimilatory sulfate-reducing bacteria. *Can. J. Microbiol.*, **19**, 375–80.

Smith, A. D. (1982). Immunofluorescence of sulphate-reducing bacteria. *Arch. Microbiol*, **133**, 118–21.

Smith, L. A., Hill, S. & Yates, M. G. (1976). Inhibition by acetylene of conventional hydrogenase in nitrogen-fixing bacteria. *Nature, Lond.*, **262**, 209–10.

Smith, R. L. & Klug, M. J. (1981). Electron donors utilized by sulfate-reducing bacteria in eutrophic lake sediments. *Appl. env. Microbiol*, **42**, 116–21.

Sørensen, J., Christensen, D. & Jørgensen, B. B. (1981). Volatile fatty acids and hydrogen as substrates for sulfate-reducing bacteria in anaerobic marine sediment. *Appl. env. Microbiol.*, **42**, 5–11.

Sorokin, Y. (1954a). New methods for isolating sulphate-reducing bacteria. *Proc. Inst. Microbiol. Acad. Sci., USSR*, **3**, 121–9.

— (1954b). Concerning the role of phosphate in chemosynthesis by sulphate-reducing bacteria. *C. r. Acad. Sci., USSR*, **95**, 661–3.

— (1957). Contribution to the question of the utilization of methane for the formation of sulphides by sulphate-reducing bacteria. *C. r. Acad. Sci., USSR*, **115**, 816–18.

— (1966a). The role of carbon-dioxide and acetate in biosynthesis in sulfate-reducing bacteria. *C. r. Acad. Sci., USSR*, **168**, 199–201.

— (1966b). Sources of energy and carbon for biosynthesis in sulfate-reducing bacteria. *Mikrobiologiya*, **35**, 761–66 (T643–7).

— (1966c). Investigation of the structural metabolism of sulfate-reducing bacteria with C^{14} *Mikrobiologiya*, **35**, 967–77 (T806–14).

— (1966d). Role of carbon dioxide and acetate in biosynthesis by sulphate-reducing bacteria. *Nature, Lond.*, **210**, 551–2.

Sparling, J. H. & Hennick, B. M. (1974). The production of hydrogen sulphide in peat. *Folia microbiol., Praha*, **19**, 59–66.

Sperber, J. I. (1958). Release of phosphate from soil minerals by hydrogen sulphide. *Nature, Lond.*, **181**, 934.

Sperry, J. F. & Wilkins, T. D. (1977). Presence of cytochrome *c* in *Desulfomonas pigra*. *J. Bact.*, **129**, 554–5.

Spiro, B. (1977). Bacterial sulphate reduction and calcite precipitation in hypersaline deposition of bituminous shales. *Nature, Lond.*, **269**, 235–7.

Spurny, M. & Dostalek, M. (1958). Bacterial release of oil: II. The influence of physical and physico-chemical relationships in oil bearing rocks. *Folia biol., Praha*, **4**, 173–83.

Stacey, M. & Barker, N. (1960). *Polysaccharides of micro-organisms*, p. 82. Oxford: Oxford Univ. Press.

Stackebrandt, E. & Woese, C. R. (1981). The evolution of prokaryotes. In *Molecular and Cellular Aspects of Microbial Evolution*, Soc. Gen. Microbiol. Symp. 32, ed. Carlile, M. J., Collins, J. F. & Mosely, B. E. B., pp. 1-32. Cambridge Univ. Press.

References

Stams, A. J. M. & Hansen, T. A. (1982). Oxygen-labile L(+) lactate dehydrogenase activity in *Desulfovibrio desulfuricans*. *FEMS Microbiol. Lett.*, **13**, 389–94.

Starka, J. (1951). Some new observations about the microbial reduction of sulphates during the formation of medicinal mud. *Biologické Listy*, **32**, 108–18 (in Czech).

Starkey, R. L. (1938). A study of spore formation and other morphological characteristics of *Vibrio desulfuricans*. *Arch. Mikrobiol.*, **8**, 268–304.

— (1958). The general physiology of the sulfate-reducing bacteria in relation to corrosion. *Producers Mon. Penn. Oil Prod. Ass.*, **22**, 12–30

— (1960/61). Sulfate-reducing bacteria – physiology and practical significance. *Lectures on Theoretical and Applied Aspects of Modern Microbiology*. Univ. Park: Univ. Maryland Press.

— (1961). Sulfate-reducing bacteria, their production of sulfide and their economic importance. *Tappi*, **44**, 493–6.

Starkey, R. L. & Wight, K. M. (1945). *Anaerobic corrosion of iron in soil*. New York: Am. Gas Assn.

Starr, M. P. & Schmidt, J. M. (1981). Prokaryote diversity. In *The Prokaryotes*, ed. Starr, M.P., Stolp, H., Trüper, H. G., Balows, A. & Schlegel, H. G. vol. 1, pp. 3–42. Berlin, Heidelberg, New York: Springer.

Steenkamp, D. J. & Peck, H. D. (1980). The association of hydrogenase and dithionite reductase activities with the nitrite reductase of *Desulfovibrio desulfuricans*. *Biochem. biophys. Res. Commun.*, **94**, 41–8.

— (1981). Proton translocation associated with nitrite respiration in *Desulfovibrio desulfuricans*. *J. biol. Chem.*, **256**, 5450–8.

Stone, R. W. & ZoBell, C. E. (1952). Bacterial aspects of the origin of petroleum. *Ind. Engng. Chem.*, **44**, 2564–7.

Stouthamer, A. H. (1977). Energetic aspects of the growth of micro-organisms. In *Microbial energetics*, ed. Haddock, B. A. & Hamilton, W. A. *Symp. Soc. gen. Microbiol.*, **27**, 285–315.

Stüven, K. (1960). Beiträge zur Physiologie und Systematik sulfatreduzierender Bakterien. *Arch. Mikrobiol.*, **35** 152–80.

Subba Rao, M. S. (1951). A biochemical study of the microbiological formation of elemental sulphur in coastal areas. Doctoral dissertation, Indian Institute of Science, Bangalore.

Suckow, R. & Schwartz, W. (1963). Redox conditions and precipitation of iron and copper in sulphureta. In *Marine microbiology* ed. Oppenheimer, C. H., pp. 187–93, Springfield, Ill.: C.C. Thomas.

Sugawara, K., Koyama, T. & Kozawa, A. (1953). Distribution of various forms of sulphur in lake-, river- and sea-muds. *J. Earth Sci.*, **1**, 17–23.

Suh, B. & Akagi, J. M. (1966). Pyruvate-carbon dioxide exchange reaction of *Desulfovibrio desulfuricans*. *J. Bact.*, **91**, 2281–5.

— (1969). Formation of thiosulfate from sulfite by *Desulfovibrio vulgaris*. *J. Bact.*, **99**, 210–15.

Suh, B., Nakatsukasa, W. & Akagi, J. M. (1968). Sulfite-reduction by *Desulfovibrio vulgaris* and *Desulfotomaculum nigrificans*. *Bact. Proc.*, P133.

Takai, Y. & Kamura, T. (1966). Mechanism of reduction in waterlogged paddy soil. *Folia microbiol., Praha*, **11**, 304–13.

Taylor, B. F. & Oremland, R. S. (1979). Depletion of adenosine triphosphate in *Desulfovibrio* by oxyanions of Group VI elements. *Current Microbiology*, **3**, 101–3.

Temple, K. L. (1964). Syngenesis of sulfide ores: an evaluation of biochemical aspects. *Econ. Geol.*, **59**, 1473–91.

Temple, K. L. & Le Roux, N. W. (1964*a*). Syngenesis of sulfide ores: sulfate-reducing bacteria and copper toxicity. *Econ. Geol.*, **59**, 271–8.

— (1964*b*). Syngenesis of sulfide ores: desorption of adsorbed metal ions and their precipitation as sulfides. *Econ. Geol.*, **59**, 647–55.

Thauer, R. K. (1982). Dissimilatory sulphate reduction with acetate as electron donor. *Phil. Trans. R. Soc. Ser. B.*, **298**, 467–71.

Thauer, R. K. & Badziong, W. (1981). Dissimilatory sulfate reduction, energetic aspects. In *Biology of inorganic nitrogen and sulfur*, ed. Bothe, H. & Trebst, A., pp. 188–98. Berlin, Heidelberg, New York: Springer-Verlag.

Thauer, R. K., Jungermann, K. & Decker, K. (1977). Energy conservation in chemotrophic anaerobic bacteria. *Bact. Rev.*, **41**, 100–80.

Thode, H. (1980). Sulphur isotope ratios in late and early precambrian sediments and their implications regarding early environments and early life. *Origins of Life*, **10**, 127–36.

Thode, H. G., Kleerekoper, H. G. & McElcheran, D. (1951). Isotope fractionation in the bacterial reduction of sulphate. *Research, Lond.*, **4**, 581.

Thomas, P. (1972). Ultrastructure de *Desulfovibrio gigas* Le Gall et de *Desulfovibrio hildenborough*. *J. Microsc., Paris.*, **13**, 349–60.

Tolokonnikova, L. I. (1977). Rate of sulfate reduction in the Azov Sea. *Mikrobiologiya*, **46**, 352–7 (T294–9).

Traore, A. S., Hatchikian, C. E., Belaich, J. P. & Le Gall, J. (1981). Microcalorimetric studies of the growth of sulfate-reducing bacteria: energetics of *Desulfovibrio vulgaris* growth. *J. Bact.*, **145**, 191–9.

Traore, A. S., Hatchikian, C. E., Le Gall, J. & Belaich, J. P. (1982). Microcalorimetric studies of the growth of sulfate-reducing bacteria: comparison of the growth parameters of some *Desulfovibrio* species. *J. Bact.*, **149**, 606–11.

Trudinger, P. A. (1970). Carbon-monoxide reacting pigment from *Desulfotomaculum nigrificans* and its possible relevance to sulfite reduction. *J. Bact.*, **104**, 158–70.

— (1976). Microbiological processes in relation to ore genesis. In *Handbook of strata-bound and stratiform ore deposits*, ed. Wolff, H. K. pp. 135–90. Amsterdam: Elsevier.

— (1982). Geological significance of sulphur oxidoreduction by bacteria. *Phil. Trans. R. Soc. Ser. B.*, **298**, 563–81.

Tsuji, K. & Yagi, T. (1980). Significance of hydrogen burst from growing cultures of *Desulfovibrio vulgaris*, Miyazaki, and the role of hydrogenase and cytochrome c_3 in energy production system. *Arch. Mikrobiol.*, **125**, 35–42.

Tudge, A. P. & Thode, H. G. (1950). Thermodynamic properties of isotopic compounds of sulphur. *Can. J. Res. (B)*, **28**, 567–78.

Tuttle, J. H., Dugan, P. R., MacMillan, C. B. & Randles, C. I. (1969). Microbial dissimilatory sulfur cycle in acid mine water. *J. Bact.*, **97**, 594–602.

Tuttle, J. H., Dugan, P. R. & Randles, C. I. (1969). Microbial sulfate reduction and its potential utility as an acid mine water abatement procedure. *Appl. Microbiol.*, **17**, 297–302.

Tuttle, J. H., Randles, C. I. & Dugan, P. R. (1968). Activity of micro-organisms in acid mine water. I. Influence of acid water on aerobic heterotrophs of a normal stream. *J. Bact.*, **95**, 1495–1503.

Ueki, A., Azuma, R. & Suto, T. (1981). Characterization of sulfate-reducing bacteria isolated from sewage digestor fluids. *J. gen. appl. Microbiol.*, **27**, 457–64.

Ueki, A. & Suto, T. (1979). Cellular fatty acid composition of sulphate-reducing bacteria. *J. gen. appl. Microbiol.*, **25**, 185–96.

— (1981). Vitamin requirement of sulfate-reducers isolated from sewage digestor fluids. *J. gen. Appl. Microbiol.*, **27**, 229–37.

Updegraff, D. W. & Wren, G. B. (1954). The release of oil from petroleum-bearing materials by sulphate-reducing bacteria. *Appl. Microbiol.*, **2**, 309–22.

Vainshtein, M. B., Matrosov, A. G., Bashunov, V. P., Zyakun, A. M. & Ivanov, M. V. (1980). Thiosulfate as an intermediate product of bacterial sulfate reduction. *Mikrobiologiya*, **49**, 855–8 (T672–5).

Vamos, R. (1971). Poisonings caused by hydrogen sulphide in sewers and inspection pits. *Munkavédelem (Hungary)*, **17**, 41–4 (in Hungarian).

Van Leeuwen, J. W., van Dijk, C., Grande, H. J. & Veeger, C. (1982). A pulse-radiolysis study of cytochrome c_3. Kinetics of the reduction of cytochrome c_3 by methyl viologen radicals and the characterisation of the redox properties of cytochrome c_3 from *Desulfovibrio vulgaris* (Hildenborough). *Eur. J. Biochem.*, **127**, 631–7.

Veldkamp, H. (1976). *Continuous culture in microbial physiology and ecology*. Patterns of Progress. Durham, UK.: Meadowfield Press.

Vernon, W. H. (1957). Metallic corrosion and conservation. In *The conservation of natural resources*, pp. 105–37. London: Inst. Civil Engn.

Vogel, H., Bruschi, M. & Le Gall, J. (1977). Phylogenetic studies of two rubredoxins from sulfate-reducing bacteria. *J. mol. Evol.*, **9**, 111–19.

von Wolzogen Kühr, C. A. H. & van der Vlugt, I. S. (1934). The graphitization of cast iron as an electrobiochemical process in anaerobic soils. *Water*, **18**, 147–65.

Vosjan, J. H. (1970). ATP generation by electron transport in *Desulfovibrio desulfuricans*. *Antonie van Leeuwenhoek*, **6**, 584–6.

— (1974). Respiration and fermentation of the sulphate-reducing bacterium *Desulfovibrio desulfuricans* in a continuous culture. *Pl. Soil*, **43**, 141–52.

Wagner, G. C., Kassner, R. J. & Kamen, M. D. (1974). Redox potentials of certain vitamins K: Implications for a role in sulfite reduction by obligately anaerobic bacteria. *Proc. natn. Acad. Sci. USA*, **71**, 253–6.

Wake, L. V., Christopher, R. K., Rickard, P. A. D., Andersen, J. E. & Ralph, B. J. (1977). A thermodynamic assessment of possible substrates for sulphate-reducing bacteria. *Aust. J. biol. Sci.*, **30**, 155–72.

Wakerley, D. S. (1979). Microbial corrosion in UK industry: a preliminary survey of the problem. *Chem Ind.*, 656–8.

Ward, D. M. & Olson, G. J. (1980). Terminal processes in the anaerobic degradation of an algal–bacterial mat in a high-sulfate hot spring. *Appl. env. Microbiol.*, **40**, 67–74.

Ware, D. A. & Postgate, J. R. (1971). Physiological and chemical properties of a reductant-activated inorganic pyrophosphatase from *Desulfovibrio desulfuricans*. *J. gen. Microbiol.*, **67**, 145–60.

Watenpaugh, K. D., Seiker, L. C., Jensen, H. L., Le Gall, J. & Dubordieu, M. (1972). Structure of the oxidized form of a flavodoxin at 2.5 Å resolution: Resolution of the phase ambiguity by anomalous scattering. *Proc. natn. Acad. Sci. USA*, **69**, 3185–8.

Watts, H. E. (1936). *Explosion in a petroleum tank at Saltend, Hull, Yorkshire*. London: HM Stationery Office.

— (1938). *Report on explosion in kerosene tank at Killingholme, Lincolnshire*. London: HM Stationery Office.

Werkman, C. H. (1929). Bacteriological studies on sulfid spoilage of canned vegetables. *Iowa agric. exp. Stn. Res. Bull.*, **117**, 163–80.

Westen, H. M. van der, Mayhew, S. G. & Veeger, C. (1978). Separation of hydrogenase from intact cells of *Desulfovibrio vulgaris*. *FEBS Lett.*, **86**, 122–6.

— (1980). Effect of growth conditions on the content and O_2-stability of hydrogenase in the anaerobic bacterium *Desulfovibrio vulgaris* (Hildenborough). *FEMS Microbiol. Lett.*, **7**, 35–9.

Widdel, F. (1980). Anaeroberabbau von Fettsäuren und Benzoesäure durch neu isolerte Arten sulfat-reduzierender Bakterien. Doctoral dissertation:Gottingen.

Widdel, F. & Pfennig, N. (1977). A new, anaerobic, sporing, acetate-oxidizing sulfate-reducing bacterium. *Arch. Microbiol.*, **112**, 119–22.

— (1981*a*). Studies on dissimilatory sulfate-reducing bacteria that decompose fatty acids. I. Isolation of new sulfate-reducing bacteria enriched with acetate from saline environments. Description of *Desulfobacter postgatei* gen. nov., sp. nov. *Arch. Microbiol*, **129**, 395–400.

— (1981*b*). Sporulation and further nutritional characteristics of *Desulfotomaculum acetoxidans*. *Arch. Microbiol.*, **129**, 401–2.

— (1982). Studies on dissimilatory sulfate-reducing bacteria that decompose fatty acids. II. Incomplete oxidation of propionate by *Desulfobulbus propionicus* gen. nov., sp. nov. *Arch. Microbiol.*, **131**, 360–5.

Wilkinson, T. (1982). An environmental programme for offshore oil operations. *Chem Ind.*, **4**, 115–23.

Wilson, P. W. & Peterson, W. H. (1931). The energetics of heterotrophic bacteria. *Chem. Rev.*, **8**, 427–80.

Windle-Taylor, E. (1962). Iron in Waltham Abbey Well. *Metrop. Wat. Bd Rep.*, **40**, 61–2.

Winogradsky, H. (1951). Etude comparée des conditions ecologiques et des conditions de laboratoire dans le cas des thiorhodacées. *Annls. Inst. Pasteur, Paris*, **81**, 441–52.

Wood, E. C. (1961). Some chemical and bacteriological aspects of East Anglian waters. *Proc. Soc. Wat. Treat. Exam.*, **10**, 76–90.

Wood, P. M. (1978). A chemiosmotic model for sulfate respiration. *FEBS Lett.*, **95**, 12–18.

Work, E. & Dewey, D. L. (1953). The distribution of α,ε-diaminopimelic acid among various micro-organisms. *J. gen. Microbiol.*, **9**, 394–409.

Xavier, A. V., Moura, J. J. G., Le Gall, J. & DerVartanian, D. V. (1979). Oxidation reduction potentials of the hemes in cytochrome c_3 from *D. gigas* in the presence and absence of ferredoxin by EPR spectroscopy. *Biochimie*, **61**, 689–95.

Yagi, T. (1958). Enzymic oxidation of carbon monoxide. *Biochim. biophys. Acta*, **30**, 194–5.

— (1959). Enzymic oxidation of carbon monoxide II. *J. Biochem., Tokyo*, **46**, 949–55.

— (1969). Formate:cytochrome oxidoreductase of *Desulfovibrio vulgaris*. *J. Biochem., Tokyo*, **66**, 473–8.

— (1970). Solubilization, purification and properties of particulate hydrogenase from *Desulfovibrio vulgaris*. *J. Biochem., Tokyo*, **68**, 649–57.

— (1979). Purification and properties of cytochrome *c*-553, an electron acceptor for formate dehydrogenase of *Desulfovibrio vulgaris*, Miyazaki. *Biochim. biophys. Acta*, **548**, 96–105.

Yagi, T., Honya, M. & Tamiya, N. (1968). Purification and properties of hydrogenases of different origins. *Biochim. biophys. Acta*, **153**, 699–705.

Yagi, T., Inokuchi, H. & Kimura, K. (1983). Cytochrome c_3, a tetrahemoprotein electron carrier found in sulfate-reducing bacteria. *Acc. Chem. Res.*, **16**, 2–7.

Yagi, T. & Tamiya, N. (1962). Enzymic oxidation of carbon monoxide III. Reversibility. *Biochim. biophys. Acta*, **65**, 508–9.

Yates, M. G. (1967). Stimulation of the phosphoroclastic system of *Desulfovibrio* by nucleotide triphosphates. *Biochem. J.*, **103**, 32–4C.

— (1969). A non-specific adenine nucleotide deaminase from *Desulfovibrio desulfuricans*. *Biochim. biophys. Acta*, **171**, 299–310.

Young, J. W. (1936). The bacterial reduction of sulphates. *Can. J. Res. (B)*, **14**, 49–54.

Zeikus, J. G., Dawson, M. A., Thompson, T. E., Ingvorsen, K. & Hatchikian, E. C. (1983). Microbial ecology of volcanic sulphidogenesis: Isolation and characterization of *Thermodesulfobacterium commune* gen. nov. and sp. nov. *J. gen. Microbiol.*, **129**, 1159–69.

Zillig, W., Stetter, K. O., Prangishvilli, D., Schäffer, W., Wunderi, S., Janekovic, D., Holz, I. & Palm. P. (1982). *Desulfurococcaceae*, the second family of the extremely thermophilic, anaerobic, sulfur-respiring *Thermoproteales*. *Zentbl. Bakt. ParasitKde Abt. Orig.*, **C3**, 304–17.

ZoBell, C. E. (1946). *Marine Microbiology*. Waltham, Mass.: Chronica Botanica Co.

— (1947a). Bacterial release of oil from oil-bearing materials. *Wld Oil*, **126**, 36–40; **127**, 35–41.

— (1947b). Microbial transformation of molecular hydrogen in marine sediments, with particular reference to petroleum. *Bull. Am. Assn. Petrol. Geol.*, **31**, 1709–51.

— (1950). Bacterial activities and the origin of oil. *Wld Oil*, **130**, 128–38.

— (1958). Ecology of sulfate-reducing bacteria. *Producers Mon. Penn. Oil Prod. Ass.*, **22**, 12–29.

ZoBell, C. E. & Morita, R. Y. (1957). Barophilic bacteria in some deep sea sediments. *J. Bact.*, **73**, 563–8.

Zubieta, J. A., Mason, R. & Postgate, J. R. (1973). A four-iron ferredoxin from *Desulfovibrio desulfuricans*. *Biochem. J.*, **133**, 851–4.

Index

acetate metabolism, 67–8
  growth characteristics, 39
  and methanogenesis, 120, 121
acetate oxidation, 5–6
  energy relations, 57, 58
acetate utilization, 67–8
  in sulfureta, 119–20
  and taxonomy, 23
adenosine monophosphate (AMP), 85, 87
  lactate oxidation, 91–2
adenosine phosphosulphate (APS)
  reduction of, 85–6, 87
adenylsulphate reductase
  location in cell, 52
Ain-ez-Zauia, 138–40
amino acids
  chelating action with iron, 44
  composition, 52–3
  sulphur containing, 4, 97–8
  transamination, 97
p-aminobenzoic acid, 44
AMP, see adenosine monophosphate
anaplerosis, 59n, 68
anatomy, 51–2
  cell walls, 53
  flagella, 11, 18, 19, 51
  spheroplasts, 52, 53–4
animal nutrition, 150–1
antibiotics, 54–5
APS, see adenosine phosphosulphate
assimilatory sulphate reduction, 1
  APS and PAPS formation, 85
  in sulphur cycle, 4
ATP
  generation, 60–1, 90–2, 94–5
  and phosphorus metabolism, 98
  sulphate reduction, 91–2
ATP-sulfurylase, 85
autotrophy
  and mixotrophy, 7–8
  partial, 47

Bacteroides, 102
barophily, 43, 109
  and barotolerance, 43, 148
Beggiatoa spp., 4, 117
Biosphere
  carbon mineralization and sulphate reduction, 122
  evolution of sulphur metabolism, 102–3
  geochemical oxidant sink, 112
  sulphur, 1; deposition, 138–40, 142–4; exploitation, 138, 140, 142
  see also sulphur cycle
biotechnology
  biochemical fuel cells, 152
  and hydrogenase, 72
  hydrogen production, 152
  and industrial wastes, 130–1; distillery, 131;
    mine waters, 131; paper industry, 131;
    sewage, 130–1, 131–2, 140–1
  metabolic regulation, 99
  methane production, 132
  sulphur production from sewage, 140–1;
    justification for, 142
biotin
  requirement for, 13n, 44
British National Collection of Industrial
  and Marine Bacteria (NCIMB), 28, 38, 155–61

Calvin cycle, 62
Campylobacter spp., 4
carbon dioxide assimilation, 61–2
carbon isotope fractionation, 70
carbon mineralization
  and sulphate reduction in sulfureta, 121–2
carbon source dissimilation, 62–70
  and taxonomy, 10, 12–13, 23
cathodic depolarization, 133–5
  protection against, 137
chelation, 44

200

Index

chemostats
 continuous culture, 40–1; and isolation, 36–7
 culture media, 31, 32
 growth yields and energy budgets, 92–3
Chlorobium spp., 115, 139
Chlorochromatium spp., 115
Chloropseudomonas ethylica, 115
 cytochrome c_7, 78
choline, 12, 42, 64
chromate, 83
 growth inhibition, 48, 84, 100, 129
Chromatium spp., 125, 128, 139
citrate, 63
 synthesis of, and evolution, 104
citrate synthase, 65, 67
citric acid cycle, 67–8
Clostridium spp., 60, 102
Clostridium nigrificans
 serology and identification, 27
 see also Desulfotomaculum nigrificans
commensalism, 111
consortia and syntrophy, 114–16
corrosion
 cathodic depolarization, 133–5
 costs of, 132, 135
 by hydrogen sulphide, 135, 136, 137–8, 150
 by iron sulphide, 135, 136
 of metals: aluminium, 136; copper, 136, 138;
  iron, 111, 127, 133–6; zinc, 136
 by non-sulphate-reducing bacteria, 136
 protection against, 136–7
 and redox potential of soil, 135–6
 stonework, 137–8
 see also spoilage
cultivation, 153
 growth characteristics, 39–42
 impurities, 4–5; and misleading results, 7, 10, 62
 isolation, 36–7, 153; of mixotrophic *D. vulgaris*, 47
 purity checks, 38
 for qualitative tests, 34–5
 for quantitive tests, 35–6
 and redox potential, 30, 59
 routine, 37–8
 salt relations, 42–3
 sample collection, 34
 storage of cultures, 38–9; freeze drying, 38–9, 49; medium from, 31, 32
 temperatures for, 34
 see also chemostats; growth; nutrition
ulture media, 30–4
 for diagnoses, 12–13, 30, 31, 32
 for mixotrophic *D. vulgaris*, 47

redox potential, 30 and n
 see also nutrition
cysteine, 64
'cytochrome c_3', 73–5
 and water synthesis, 74
cytochromes
 b, 21, 78; and electron transport, 82
 b-type, 29, 74, 78; location in cell, 52
 c, 21; and iron metabolism, 98–9
 c_3, 21, 55, 73; and n, 74, 75–6; and carbon metabolism, 69; and electron transport, 81–2; and ferredoxin, 80; and iron metabolism, 98–9; and Knallgass reaction, 70; and pyruvate metabolism, 67
 cc_3, 74, 76, 77; and electron transport, 82
 c_7, 78
 c_{553}, 73n, 74, 77; and electron transport, 81; folding pattern and evolution, 104
 c-type, 29; and location in cell, 52
 d, 78; and electron transport, 73–8, 81–2;
 and evolution, 104; nomenclature, 73n; properties of, 73–5; and taxonomy, 11, 21
d-type, 74
 Yagi's, 77

dehydrogenase, 63
Desulfobacter spp.
 pyruvate metabolism, 66, 67
Desulfobacter postgatei, viin, 16
 acetate utilization, 67–8, 119, 121
 cultivation: growth, 39; nutrition, 12, 33–4, 44
 cytochromes: *b*, 78; *c*, 75
 diagnostic characters, 11, 12
 in marine sediments, 114
 morphology, 11, 16, 20; ultrastructure, 52
 motility, 11, 21
 in sulphur cycle, 4
Desulfobulbus spp.
 in marine sediments, 114
 motility, 21
Desulfobulbus propionicus
 cytochromes: *b*, 78; *c*, 75
 diagnostic characters, 11, 12, 23
 lactate metabolism, 64
 nitrate reduction, 96
 nutrition, 12, 44; sulphate-free growth, 67
Desulfococcus spp.
 desulfoviridin and phylogeny, 101

Desulfococcus multivorans
diagnostic characters, 11, 12, 23
morphology, 11, 16, 20
Desulfomonas pigra, 28–9
Desulfonema spp.
acetate utilization, 119
autotrophy, 8
diagnostic metabolic processes, 23
motility, 11, 21
phylogeny, 101
tactophily, 42
Desulfonema limicola
desulfoviridin, 11, 22; and phylogeny, 101
diagnostic characters, 11, 13
morphology, 11, 16, 20
Desulfonema magnum
diagnostic characters, 11, 13
morphology, 11, 16, 20
desulforedoxin, 80
Desulforistella spp., 28
desulforubidin, 89
Desulfosarcina spp.
autotrophy, 8
diagnostic metabolic processes, 23
light sensitivity, 34, 49–50
Desulfosarcina variabilis
and desulfoviridin, 22
diagnostic characters, 11, 13
light sensitivity, 49–50
morphology, 11, 16, 20
Desulfotomaculum spp.
characteristics, 9, 18–20
cultivation, 37–8; anaerobic atmosphere, 37;
growth, 39; media, 31, 32; storage, 38–9
cytochrome *b*, 21, 78
and economic activities: food spoilage, 145;
metal corrosion, 133
ferredoxins, 78
habitats, 109
and microbicides, 47
nitrogen fixation, 44
Ouchterlony plate, 26
pyrophosphate utilization, 94
pyruvate metabolism, 66–7
qualitative test, 34–5
saline acclimatization, 109
sporulation in, 11, 18, 20; medium for, 20
in sulphur cycle, 4
sulphate-free growth, 41–2, 67
sulphite reductase, 86
Desulfotomaculum acetoxidans
acetate: oxidation, 5, 68; utilization, 23, 119

cytochrome c_7, 78
diagnostic characters, 11, 13
habitats, 20, 109, 113, 151
morphology, 17
nutrition, 13, 44
sporulation, 11, 17, 20
temperature relations, 23, 34
Desulfotomaculum antarcticum
carbon source utilization, 13; sugar, 23
diagnostic characters, 11, 13
Desulfotomaculum nigrificans
ATP-sulfurylase, 85
carbon source utilization, 13; glucose, 13, 65
diagnosis: characteristics for, 11, 13; medium for, 31, 33
economic activities: corrosion, 136; food spoilage, 144–5
Gram reaction, 21
hibitane tolerance, 25
morphology, 19; pleomorphism, 14, 17
NCIMB strains, 156
phylogeny, 101
spore heat resistance, 20, 49
sulphite reductase, 89
thermophily, 6–7, 9, 109
α-tocopherol, 55
Desulfotomaculum nigrificans subsp.
salinum, 27
Desulfotomaculum orientis
diagnostic characters, 11, 13, 18
hibitane tolerance, 25
spore heat resistance, 49
Desulfotomaculum ruminis
diagnostic characters, 11, 13
habitat, 113, 151
hibitane tolerance, 25
hydrogenase, 73
nitrogen fixation, 151
nutrition, 13, 44
spore heat resistance, 49
temperature relations, 23
Desulfovibrio spp.
barophily and barotolerance, 43, 108, 148
carbon dioxide assimilation, 61–2
cultivation, 37–8; continuous culture, 40–1,
(alkalinity and salt precipitation), 41;
media, 31, 32; qualitative and quantitative tests, 34–6; and storage, 38–9
cytochromes, 21, 73–8; and electron transport, 81–2
destruction: by heat, 48–9; by microbicides, 47–8
desulfoviridin, 11, 22, 88

Index 203

Desulfovibrio (contd.)
  diagnostic characters, 9, 10
  dissimilatory sulphate reduction, 60–1
  and economic activities: metal corrosion, 133;
    oil technology, 147, 148; paper technology, 112, 113, 150
  enthalpy data, 56
  enzymes: ATP-sulfurylase, 85; localization of,
    52; sulphite reductase, 86; superoxide dismutase, 104
  evolution, 103–4; and DNA content, 101
  growth: 39; on 'incomplete' substrates, 45–7;
    sulphate-free, 41–2, 67, 69
  impermeability to sulphate, 84 and n
  metabolism: acetate, 28, 64; formate, 64–5;
    hydrogen, 70–3; iron, 98–9; lactate, 63–4; phosphorus, 98; products of, 68–9; pyruvate, 66–7
  mineral deposition: native sulphur, 138–9;
    soda, 144
  mixotrophy, 7–8, 47
  morphology, 11, 14, 18; pleomorphism, 14, 17,
    (and sulphide concentration) 16, 18
  motility, 11, 21
  nitrogen fixation, 27–8, 44, 97, 111
  pathogenicity, 151
  populations and habitats, 109;
  quantitative data, 112–4
  synonyms, 9
  syntrophy, 115
  tactophily, 42
  thermophily, 109
Desulfovibrio africanus
  cytochromes: b, 78; c_3, 75; d, 78
  diagnostic characters, 11, 12
  ferredoxins, 79
  hibitane tolerance, 25
  molybdoproteins, 55
  rubredoxins, 80
Desulfovibrio baarsii
  acetate utilization, 23, 119
  diagnostic characters, 11, 12
Desulfovibrio baculatus, 29
  diagnostic characters, 11, 12
  morphology, 14, 15
Desulfovibrio desulfuricans
  amino acid composition, 52–3
  cell walls, 53
  cytochromes: c_3, 75; cc_3, 77
  diagnostic characters, 11, 12, 18

fumarase, 66
  growth on 'incomplete' substrates, 45–6
  hibitane tolerance, 25
  hydrocarbons, 54; formation, 69
  menaquinones, 81
  mucopolysaccharides, 54
  nitrogenase activity, 97
  permeability to ammonium ions, 96
  rubredoxins, 80
  salt tolerance, 42, 43
  spheroplasts, 53–4
  sulphur amino acid utilization, 97–8
  syntrophy, 115, 116
Desulfovibrio desulfuricans, Berre strains, 14, 18, 28
Desulfovibrio desulfuricans, strain E1 Agheila, 139
Desulfovibrio desulfuricans, strain Essex, 15
Desulfovibrio desulfuricans, strain Norway, 4
  catalases, 55
  cobalt-containing proteins, 55
  desulforubidin, 89
  and desulfoviridin, 11, 22
  ferredoxins, 79
  morphology, 14
  superoxide dismutase, 55
Desulfovibrio desulfuricans subsp. aestuarii, 27, 157
Desulfovibrio gigas
  cobalt-containing proteins, 55
  cytochromes: b, 82; c_3, 74, 75, 76; cc_3 77; c-type, 52
  desulforedoxin, 80
  diagnostic characters, 11, 12
  enzymes: catalase, 55; hydrogenase, 72, (location) 52, 72; lactate dehydrogenase, 63–4; superoxide dismutase, 55
  evolution, 103
  ferredoxins, 79–80, 92
  flavodoxins, 81
  fumarate metabolism, 66
  hibitane tolerance, 25
  hydrogen cycling, 95
  menaquinones, 81
  molybdoproteins, 55
  morphology, 14, 15, 17; plasmids, 101; ultrastructure, 51–2
  respiratory chain phosphorylation, 92
  rubredoxins, 80
  spheroplasts, 54
'Desulfovibrio hildenborough', 28
Desulfovibrio rubentschickii, 5
Desulfovibrio salexigens
  cytochromes: c_3, 75; c_{553}, 77
  diagnostic characters, 11, 12

Desulfovibrio salexigens (contd.)
 flavodoxins, 81
 halophily, 42–3
 microbicide resistance: hibitane, 25;
  quaternaries, 128
 pleomorphism, 14
 rubredoxins, 80
Desulfovibrio sapovorans
 diagnostic characters, 11, 12
 fatty acid utilization, 45, 68
 morphology, 15
Desulfovibrio thermophilus
 diagnostic characters, 11, 12
 thermophily, 23
Desulfovibrio vulgaris
 anatomy, 51–2
 carboxylic acid cycle, 65
 cell walls, 53
 cytochromes: c_3, 75, 76; c_{553}, 77; cc_3, 77; c-type, 52
 desulfoviridin, 88
 diagnostic characters, 11, 12
 dye reduction tests, 71
 enzymes: APS-reductase, 86; catalase, 55;
  hydrogenase, 72, 152, (location) 52;
  lactate dehydrogenase, 63; nitrite reductase, 96; superoxide dismutase, 55
 evolution, 103
 ferredoxins, 78–9
 flavodoxins, 81
 hibitane tolerance, 25
 iron uptake, 44
 menaquinones, 81
 and methanogens, 121
 mixotrophy, 47
 mucopolysaccharides, 54
 mytomycin C effect, 52
 oxygen tolerance, 109
 pleomorphism, 18
 pyruvate metabolism, 67
 rubredoxins, 80
 salt sensitivity, 42, 43
 spheroplasts, 53–4
 syntrophy, 115
 tactophily, 42
Desulfovibrio vulgaris, strain Hildenborourgh
 cytochrome c_3, 73n
 diaminopimelic acid, 53
 heat resistance, 49
 iron precipitation, 51
 morphology, 15; variation, 10
 and ultraviolet radiation, 50
Desulfovibrio vulgaris, strain Marburg, 47, 62

Desulfovibrio vulgaris, strain Wandle, 51
Desulfovibrio vulgaris subsp. *aestuarii*, 27
desulfoviridin, 88–9
 and evolution, 101, 104
 location in cell, 52
desulfoviridin test, 22
Desulfurococcus spp., 3, 4
Desulfuromonas spp., 4
Desulfuromonas acetoxidans
 acetate oxidation, 5
 citric acid cycle, 68
 in consortium, 115
 rubredoxin, 80
Deutsche Sammlung von Mikroorganismen (DSM), 161–2
diaminopimelic acid, 53
dissimilatory sulphate reduction
 catabolism, 60, 61
 defined, 1–2
 in sulphur cycle, 4
dissimilatory sulphur metabolism, *see under*
 metabolism
 distribution, 108–9
 correspondence with ecology, 107
 in natural environments: quantitative data, 112–14
 and redox potential, 109
dithionite
 reduction, 74, 86, 87, 90–1
DNA
 and taxonomy, 11, 22
 and evolution, 101
DNA–rRNA homology
 and numerics, 25
DSM (Deutsche Sammlung von Mikroorganismen), 161–2

ecology
 correspondence with distribution, 107
 and population size, 107
 of sulfureta, 119
 and sulphur isotopes, 108
 see also environment
EDTA, *see* ethylenediaminetetra-acetic acid
electron transfer
 energy relations, 58–9
 redox potential, 58–9
electron transport
 catabolism, 60–1
 cytochromes, 73–8
 and ferredoxins, 78–80
 mechanisms, 81–2
 energy relations, 56–9
 acetate oxidation, 58
 electron transfer, 58–9

energy relations (contd.)
  hydrocarbon oxidation, 58
  methane utilization, 58
  phosphorus metabolism, 98
  redox potential, 58–9
  sulphide hydrolysis, 56, 58
  in syntrophy, 116
  see also metabolism
Entner–Doudoroff pathway, 65
environment
  effects of sulphate reduction: commensal flora alteration, 111; geochemical oxidant sink, 112; heavy metal removal, 110–11; hydrogen removal, 111; isotope fractionation, 112; nitrogen fixation, 111; sulphate removal, 111; sulphide formation, 110 and n; organic matter removal, 111; pH alteration, 110
  of sulfureta, 119
  sulphate-deficiency development, 132
  see also ecology
enzyme preparation
  by freezing, 49
  by mechanical and sonic damage, 49
ethylenediaminetetra-acetic acid (EDTA), 53–4, 168
evolution
  and aerobic metabolism, 105–6
  and gene transfer, 105
  and primitive properties, 103–5
  of sulphur metabolism, 102–3
  trophism, 105

fatty acids
  in metabolism, 68
  requirements and taxonomy, 12, 21, 23
fermentative growth, 2
ferredoxins
  characteristics, 78–80
  location in cell, 52
  in metabolism: and ATP generation, 92; citric acid cycle, 68; and cytochrome c_3, 76; and electron transport, 78–80; and evolution, 103, 104; of pyruvate, 47, 62, 67
  structure and activity, 79–80
flagella
  morphology, 11, 18, 19, 51
  taxonomy, 11, 18
flavodoxin, 81, 82
  and cytochrome c_3, 76
food spoilage, 144–5
formate, 45, 57
  metabolism, 64–5, 70
fossil formation, 144

freeze-drying, 38–9, 49
fumarate, 12–13, 45, 57, 66
  respiratory chain phosphorylation, 92
  and sulphate-free growth, 42, 69
fumarate reductase
  location in cell, 52

gas contamination, 150
Gram reaction, 20–1
growth
  sulphate-free, 41–2, 67, 69–70
  test medium, 31
  see also cultivation
guanine: cytosine ratio, 6, 11, 22, 29

halophily, see salt, tolerance
hibitane, 167
  resistance to, 25
hydrocarbons
  co-oxidation, 148
  energy relations of oxidation, 58
  and evolution, 104
  formation, 54, 145–6
  metabolism, 68, 69
  see also methane
hydrogen, 66
  in chemotrophic assimilation, 47
  substrate for sulphate reduction, 120
hydrogen cycling, 95
hydrogenases, 70–3
  activity: cell-free, 71–2; dye tests for, 70–1
  forms of, 72
  location in cell, 52, 72–3
  oxygen sensitivity, 72, 99
  purification of, 71
  synthesis, 72
hydrogen scavenging, 63, 64, 111, 115, 121
hydrogen sulphide
  catalyst poison, 37
  corrosion by, 135, 136, 137–8, 150
  gas contaminant, 150
  and growth in culture, 39
  from industrial wastes, 132
  neutralization of inhibitors, 48
  oceanic geothermal, and sulfureta, 118
  in rice paddies, 130
  from sand pollution, 129
  from sewage sludge, 140–1
  spoilage by, 124–6
  toxicity, 40 and n, 125, 126, 143

industrial wastes
  treatment: distillery slop, 131; mine waters, 131; paper industry liquor, 131; sewage, 130–2;
  see also pollution

206 Index

inhibition, 47–8, 128–9,
  by aeration, 48, 127–8
  chemical, 48, 99–100, 128–9; chromate, 48, 84, 100, 129; cyanide, 100; dyes, 100, 128; fluoracetate, 120; hibitane, 25; molybdate, 48, 84, 120, 121; nitrates, 128–9, 130, 150; quaternaries, 128, 164; selenate, 48, 83, 84, 87, 100; sodium azide 99–100
  inhibitors listed, 163–8
  by visible light, 49–50
  see also killing
iron
  and chelation, 44
  deficiency, 51, 110–11, 113
  metabolism, 98–9
  precipitation, 51, 110–11

killing
  methods: cold, 49; desiccation, 49; heat, 48–9; mechanical damage, 49; sonic damage, 49; ultraviolet radiation, 50
  see also inhibition
Klebsiella aerogenes, 2
Knallgassbakterien (hydrogen bacteria), 61
Knallgass reaction, 61, 70

lactate, 12–13, 57, 66
  oxidation of, 63–4, 91, 120
  and sulphate-free growth, 69, 70, 115
light sensitivity, 34, 49–50

malate, 12–13, 45, 57, 63, 66
  and sulphate-free growth, 42, 69
marsh gas, 117
menaquinones, 81
metabolism
  anabolism, 59–60
  carbohydrate utilization, 65
  carbon dioxide assimilation, 61–2
  carbon dissimilation, 62–70
  catabolism, 59–61
  dissimilatory sulphur metabolism, 82–91;
    activation of sulphate, 85, 87;
    adenosine phosphosulphate reduction, 85–6; and analogues of sulphate ion, 83–4; cyclic pathway, 87–8; desulfoviridin, 88–9; impermeability to sulphate, 84 and n; inhibitors, 84; sulphide formation, 84, 87; sulphite as intermediate, 82–3; sulphite reduction, 86–9; sulphur isotope fractionation, 89–90
  electron transport, 73–82
  energy balance, 91–6; and chemiosmosis, 94–5;
  and growth yields, 92–4; and fumarate, 92; and hydrogen, 93–4; hydrogen cycling, 95; lactate oxidation and sulphate reduction, 91–2; and pyruvate, 93; respiratory chain phosphorylation, 92–4, 94–5
  evolution, 102–3
  growth substances: acetate, 67–8; choline 64; cysteine, 64; fatty acids, 68; formate, 64–5; fumarate, 66; hydrocarbons 68, 69; pyruvate, 64, 66–7
  hydrogen, 70–73
  iron, 8–9
  lactate dehydrogenases, 63–4
  methane production, 69
  nitrogen, 96–8; nitrate utilization, 96
  phosphorus, 98
  products of, 68–9
  regulation of, 99
  sulphate-free, 69–70
  sulphide inhibition, 93–4
  thermodynamics of reactions, 56–9
  unsolved mysteries, 153–4
methane
  formation, 69
  from sewage treatment, 132
  oxidation of, 122
  and sulfureta, 117
  and vitamin B_{12}, 67
  see also hydrocarbons
Methanobrevibacter smithii
  syntrophy, 115
methanogenesis
  and hydrogen scavenging, 121
  in sulfuretum development, 117
  and sulphate reduction, 120–1
  and sulphide, 120–1
Methanosarcinia barkeri
  acetate utilization, 121
  syntrophy, 115
mineral deposition
  metal sulphide ores, 142–4
  soda, 143, 144
  sulphur, 138–42
mitomycin C, 52
mixotrophy, 7–8, 45–7
  acetate assimilation, 23
  carbon dioxide assimilation, 61–2
  continuous culture, 41
  and evolution, 103–4
molybdate
  inhibition by, 48, 84, 120, 121
monofluorophosphate, 83, 84
morphology, 11, 14–20
motility, 11, 21
mucopolysaccharides, 54, 69

NCIMB, (British National Collection of Industrial and Marine Bacteria), 28, 38, 155–61
nitrates
  as inhibitors, 128–9, 130, 150
  reduction of, 96; and chemiosmosis, 95
nitrilotriacetic acid, 41
nitrite reductase, 96
nitrogenase, 97
nitrogen fixation, 27–8, 44, 96–7, 111
  marine, 122–3
nitrogen metabolism, 96–8
  organic carbon utilization, 97–8; and sulphur amino acids, 97–8
nutrition, 44–5
  see also culture media

oil technology, 145–9
  bacteria in oil deposits?, 146
  bacterial release of oil, 146–7
  contamination of injection waters, 148
  co-oxidation of hydrocarbons, 147–8
  spoilage of stored products, 148–9; 'cutting emulsions', 149; explosive iron sulphide, 149; petroleum, 148–9
Ouchterlony plates, 24, 26

paper technology, 150
  bacteria in mill effluent, 112, 113
PAPS (phospho-adenosinephosphosulphate), 85
pathogenicity, 151
Pelochromatium spp., 115
Peptococcus aerogenes, 79
pH
  pollution treatment, 128
  and sulphate reduction, 110
  tolerance, 108
phosphate release
  phytoplankton production, 129
phospho-adenosinephosphosulphate (PAPS), 85
phosphorus metabolism
  and energy transfer, 98
pleomorphism, 14, 16, 17, 18
  poisonous dawn fog, 126
pollution
  and gas storage, 150
  prevention and treatment, 48, 127–9
  of sand and soil, 129–30
  of water, 124–9; fish death, 125–6; hydrogen sulphide, 124–5; phosphate release and phytoplankton production, 129
  see also industrial wastes

populations in natural environments, 107, 112–14
primitive properties, 103–5
Prosthecochloris aestuarii, 115
psychrophily, 44, 49, 108
pyrophosphatase, 94, 99
pyrophosphate, 85, 87
pyruvate, 12–13, 45, 57
  metabolism of, 64, 66–7
  and mucin, 54
  and sulphate-free growth, 41–2, 64, 67, 69
pyruvic phosphoroclasm, 67, 82
  and evolution, 104
  and ferredoxins, 80
  and flavodoxins, 81

redox-poising agents, 30
redox potential
  and distribution, 109; in water, 113
  electron transfer, 58–9
  energy relations, 58–9
  and growth, 30 and n, 59
  soil corrosiveness, 135–6
respiratory chain phosphorylation, 80, 92–4, 94–5
rRNA
  evolution and relatedness, 25, 104
rubredoxin
  characteristics, 80–1
  and cytochrome c_3, 76
  and evolution, 103

salt
  concentration and pleomorphism, 16
  requirements and taxonomy, 12–13, 23, 27
  tolerance, 2, 42–3, 108, 109; 'training', 42, 43
selenate
  inhibition by, 48, 83, 84, 87, 100
  sulphate analogue, 83
serology
  Ouchterlony plates, 24, 26
  taxonomy, 24–7
sewage treatment, 130–2
  methane production, 132
  sulphide/sulphur production, 140–1
sirohydrochlorin, 88–9, 104
soda deposition, 144
Sorokin's reaction, 104
spheroplasts, 52, 53–4
spoilage
  food, 144–5
  by hydrogen sulphide, 124–5, 130
  leather, 151
  see also corrosion

spores, 17
  formation of: and salinity, 42; and
    taxonomy, 11, 18, 20
  heat resistance, 49
Sporovibrio ferro-oxidans, 9n
sulfureta, 116–23, 140
  acetate-utilization, 119–20
  development, 4, 111, 116–17, 118
  laboratory study, 119
  metal sulphide ores, 142–4
  methane oxidation, 122
  nitrogen fixation, 122–3
  organic carbon mineralization, 121–2
  and oxygen in atmosphere, 112
  sulphate reduction and methanogenisis, 120–1
sulphide
  formation, 86, 110
  hydrolysis, 56, 58
  and methanogenisis, 120–1
  oxidation, 4
sulphide ore formation, 142–4
sulphite
  and growth, 41, 69
  intermediate in sulphate reduction, 82–3
  paper industry waste, 131, 150
  reductase system, 86–9; products from, 86–8;
    and cytochrome c_3, 74, 76
  respiratory chain phsophorylation, 92
sulphur
  deposition, 138–40, 143–4
  exploitation, 140, 142
  microbial production, 140–1; justification for, 142
  reduction 41, 91
sulphur cycle, 1, 3–4
  see also biosphere
sulphur isotope fractionation, 102–3, 112
  and sulphate reduction, 89–90
  thermodynamics, 90
sulphur-reducing bacteria, 41
superoxide dismutase, 55, 104
Syntrophobacter wolinii, 115, 116
Syntrophomonas wolfii, 115
syntrophy and consortia, 114–16

tactophily, 37, 42
taxonomy
  and acetate utilization, 23
  carbon source utilization, 12–13, 23
  and chemical composition, 21
  and culture impurity, 10
  cytochromes, 21
  desulfoviridin, 11, 22
  dissimilatory sulphate reduction, 10
  and DNA composition, 11, 22
  DNA–rRNA homology, 25

Gram reaction, 20–1
guanine : cytosine ratios, 6, 11, 22, 29
and hibitane resistance, 25
morphology, 14–20
motility, 11, 21
and salt relations, 23, 27
serology, 24–7
strain designation, 28
and thermophily, 12–13, 23
temperature
  and cultivation, 34
  specificity of, 6–7
  tolerance of extremes, 108–9
  see also thermophily
tetrathionate
  and growth, 41, 69
  reduction, 74, 86, 87, 90; and cytochrome c_3, 74, 76
  in sulphite reduction, 86–7
Thermodesulfobacterium commune, 29
thermophily, 2, 108, 109
  and cultivation, 39; isolation, 36
  and taxonomy, 12–13, 23
  see also temperature
Thiobacillus spp., 128
  APS reductase, 86
  and corrosion, 137, 138
  in sulfureta, 117
  in sulphur cycle, 4
Thiomicrospira pelophila, 128
Thiopedia spp., 125
thiosulphate
  and growth, 41, 69, 94
  reduction, 74, 86, 87, 90; and cytochrome c_3, 74, 76
  in sulphate reduction, 86–7
Thiovulum spp.
  in sulfureta, 117
a-tocopherol, 55
tricarboxylic acid cycle, 65–6
trithionate
  reduction, 90
  in sulphite reduction, 86–7

ultraviolet radiation
  sensitivity to, 50

vitamins
  B_{12}, 67, 88
  requirements, 12, 13, 44

Walvis Bay
  water pollution, 125

yeast
  in sulphur cycle, 1, 4
  yeast extract, 32, 63
    and growth, 44, 45, 46